GARDEN
COMPOST

A Plum Tree Classic

GARDENING WITH COMPOST

by

F. C. King

Plum Tree

This Edition Published In 2000 By
Plum Tree Publishing Limited

First published by Faber & Faber 1944

Copyright: F. C. King
& Plum Tree Publishing Limited

Artistic Illustrations by Tina Thornton

*Plum Tree Publishing Ltd
Worcestershire
England WR7 4NX*

Printed and bound in Great Britain

ISBN : 0 95336461 5

The original text of this book was written over fifty years ago. In light of this, certain methods and techniques may have changed. It is essential, therefore, before following any of the suggestions contained herein, that the reader obtains current, up to date advice. The text of this book is an abridged version of the original.

CONTENTS

Preface

This interesting and addictive little book is very much a gardening classic. The author, F. C. King was one of the early pioneers of organic gardening and his principles, along with those of several others at that time, formed the framework of organic growing as we know it today.

The original book was written during the latter part of the second world war, when good food was in short supply and there was a definite advantage in growing fruit and vegetables in a self-sufficient manner. Mr King's encouragement to produce crops without the use of artificial substances must have seemed unique at the time. He had an overwhelming desire to persuade all gardeners to follow a natural and non-chemical system. His forecast about the use of chemical fertilisers still rings true today.

The author's career spanned over forty years, of which more than thirty of these were spent as Head Gardener at Levens Hall, a magnificent Elizabethan Mansion in South Cumbria. Today Levens Hall is famous for its superb topiary gardens and is open to the public during the months of April to October.

Not only is the book a guide to producing good compost and using it efficiently, but it also covers a variety of allied subjects. These include the principles of good gardening, soils, manures, worms, weeds and many other hints and tips that help towards better results. The book is also a small insight into the history of the time and in light of this, references to costs in the examples given have been left in 'old money'. Towards the end of the book we have included the original introduction by Sir Albert Howard. Mr King also wrote a small booklet called 'Is Digging Necessary?' and we have included this as chapter 12.

During the early stages of re-publishing this book we were in contact with Mr King's daughter-in-law, Mrs Margaret King. We can think of no better way of starting the text than by including a copy of a letter sent to us, in which Mrs King gives a brief summary of the author's career.

Dear Sir

Mr King was born in Little Ribston, a small village in the North Riding of Yorkshire. When he left school he became apprenticed to a well known nursery garden, *Backhouses Of York*, from there, the 1914-18 war having started, he joined *The Manchester Pals* regiment, was wounded at The Somme and invalided out. After repatriation he became employed as a gardener at *Lowther Castle*, the estate of The Earls of Lonsdale, near Penrith.

He took up employment as Head Gardener at Levens Hall in 1919, and was there until his retirement in 1955, when he and my husband (his son) set up a Market Garden in Hale, a small village only 10 minutes away from Levens Hall, and he lived there until his death in 1975. My husband passed away in 1990.

Mr King's father was a stone mason on the Ribston estate, which is now defunct.

I am enclosing the little pamphlet 'Is Digging Necessary?' also a hand written article which I have found, I think that this may have been written in reply to an article which appeared in The Westmorland Gazette at the time, I hope you will find these of interest.

Just in passing, he was an accomplished cricketer - being a Yorkshire man - what else! also a very good fly fisherman.

I look forward to hearing from you.

Yours sincerely

Margaret King

Chapter 1

THE PRINCIPLES OF GOOD GARDENING

'Nature is one and indivisible': I would like to see this written in large letters (metaphorically at least, if not actually) over every garden gate, for many of the problems which face the gardener and in particular the amateur, would assume much less serious proportions if he saw them against the background of the whole of nature's law. The ground is prepared; the seed is sown and germinates; the seedling appears; the plant develops, flowers, bears fruit, ripens, withers, decays, and returns to the earth from where it sprang. This is nature's round, in which every happening in the garden has its place: every success or failure should show us how closely or how little we have observed and followed the order laid down by mother earth. To grasp this fundamental conception of gardening, is to have a firm anchor amid the drifting currents of theories, opinions and advice which are poured upon the inexperienced gardener from all sides.

The Foundation of Soil Fertility

Let us then take first things first. We must begin with the soil and the first essential in all gardens is soil fertility. The full value of any crop can only be secured from healthy, fertile soil, which must be treated as a living medium and not, as is too often the case, as a dead mass which can best be induced to produce a crop by repeated doses of highly concentrated so-called plant foods or artificial fertilisers. These not only rob the soil of its true fertility, but leave a legacy of pests and diseases as a measure of their efficiency. True fertility depends on humus and to create an adequate and continuous supply of humus we must see that all the waste products of the garden are returned to the soil.

Nature's round is a true circle: it does not begin or end. The soil which feeds the plants must itself have food manufactured from the remains of those plants after they have completed their life's work of bearing fruit. And nature, in her unity and completeness, has herself provided a labour force to turn the waste products of the garden into food for the soil. One of her most valuable workmen, which we can see with the naked eye, is the earthworm, but there are millions more - chiefly moulds and microbes - which are quite invisible except under the microscope and, therefore, tend sometimes to be forgotten. When we co-operate with nature by arranging for the plant residues of the garden to be converted into humus, we help to feed the vast unseen labour force in the soil, whose waste products, in turn, feed the crop. As our crops increase with this good nourishment, there are more vegetable residues for converting into humus, and as the fertility of the soil is built up, pests and diseases disappear, because once the soil is fertile the crops protect themselves, just as the wildings of our woods and hedgerows do. Nature is one and indivisible, and the problems of the garden, such as disease or the running out of varieties, are only the by-products of some interruption in nature's round or some attempt at a short cut, which results in a greater or lesser degree of starvation in the soil.

Commercial Shortcomings

In contrast to the commercial producers of seeds and plants, who are naturally influenced by the expectation of financial reward for their labours, nature does not approve of short cuts or quick-return methods, she takes the long view: commerce thinks only of the profit of the moment, and any vegetable which is an improvement on existing types is propagated as rapidly as possible, so that in the shortest time a large quantity of seed or saleable plants can be put on the market. Commerce, therefore, as a rule, instead of concentrating on making the soil rich in humus, employs the easiest possible method of raising plants - by means of the artificial manure bag. Thus the stamina of such strains is often undermined before they reach the general public, and commercial motives encourage the unwitting gardener, with the aid of some new wash or powder, to keep his struggling plants in existence for a few weary years. This is a losing battle from the start, because there has been a departure from the natural treatment of the soil and plant - a digression from nature's round. By this I would imply that with the aid of chemical manures and sprays, plants may be induced to exist on an artificial diet, by artificial methods of cultivation and protection for an artificial length of time. This at no time represents more than it indicates, namely, that the manipulative skill of the scientist alone, makes this achievement possible and it does not portray true plant life. In contrast with nature's round this is not a true circle: the beginning is to be seen as the pressing into service of dead inorganic matter for the purpose of promoting plant growth, and the process comes to an end with the matured growth, which with its lack of quality, contrasts sharply with the performance of nature. On the debit side, also, must be added the deleterious effect on the soil due to the poisonous, insoluble residues left behind by synthetic plant food.

Nature's Way

On the other hand, nature - 'the supreme gardener' - if left to follow her own laws, sees to it that a healthy stock of all plants, suitable for the particular area in which they have become established, is available. And if, as must be inevitable in an imperfect world, certain plants appear from time to time which are weak or inappropriate to their locality, nature removes them by her own agents - pests and disease. Yet such drastic measures are far less common among the wildings of the field and forest than they are in our gardens, where man substitutes his own ideas for the laws of mother earth and then tries to counteract the ill effects of his action by attempting, with sprays and powders, or with fire, to obliterate the pests and diseases which are merely nature's censors.

I shall make no apology for repeatedly stressing in the subsequent chapters of this book the importance of soil fertility, for without this all our labour is wasted. To maintain it is our first duty; when we undertake it correctly the foundation of success is well and truly laid. The failure of a crop in ground well stocked with humus, can then be traced to unsuitability of variety or to wrong position. A plant which requires sunshine cannot be a success in a shady position, nor can a shade-loving subject survive if exposed to full sun. It has been my experience that the best way to build up soil fertility, and to keep the soil in good heart, is by means of using compost.

I have used many tons of compost and have been impressed by the almost total absence of pests such as wireworms and slugs. I do not believe it would be to our advantage to rid the garden entirely of slugs, for the good they do far outweighs the damage caused to seedling plants. That the slug feeds on our plants I freely admit, but by its death and subsequent decomposition, it returns to the soil all it borrowed during its lifetime. It is far less an enemy to the garden than those who burn or otherwise destroy the waste material from crops and therefore deprive the soil of its natural form of nourishment.

To feed the soil adequately and correctly, is without doubt our primary duty, and to fulfil this duty is to win more than half the battle.

Good Cultivation

Good cultivation must be the next stage, and a soil rich in humus will be found much more easily workable than one which has been treated with artificial manures.   In humus-rich soil, moreover, the earthworm will assist the spade in keeping the soil open; well drained and well aerated.

This is the way to give seeds and young plants the best possible chance and even if, at the outset, the only seeds and seedlings available come from weakened stock, the cumulative effect of compost in the garden, must eventually help to build up better and more disease-resistant strains. The incidence of pests and diseases will become negligible, and most of the gardening problems which today loom large in the cultivator's mind, will fall into their rightful place against the background of nature's laws.

Let the test of every piece of advice offered to you, whether by radio, by pamphlet or periodical, or by the mouth of a friend, be: "Does this do good or harm to nature's labour force and the living soil? Does this accord with the laws of mother earth?"   Unless the advice offered is based upon nature's laws, we run a grave risk in following it. Too often advice given to the amateur is a mixture of fact and fiction with the latter predominating.

Clean Cultivation

If we examine the fetish of clean cultivation and the segregation of crops, coupled with artificial feeding, we shall soon see clearly why such methods prove costly and inefficient. The lowered vitality of plants and the visitation of pests and disease may well be traceable directly or indirectly to this system, for it is never practised by mother earth. Much of the trouble we experience in the growing of our vegetable crops may be related to the practice of depriving the ground of its weed crop, thereby exposing the bare soil to the action of strong sunlight. If we persist in carrying out this mistaken idea that our plants must never suffer competition from weeds, all we succeed in accomplishing by clean cultivation is a reduction in the activity of the soil bacteria, a deterioration in the texture of the soil, and a check in the progress of our cherished plants.

I am convinced, after years of observation, that crops would give heavier yields if we were not so keen on depriving the soil of waste vegetable matter by the removal of weeds and the poisoning of the soil by the use of fungicides and insecticides. During the late summer and autumn every effort should be made to secure a good crop of weeds, for without this green carpet much of the fertility of the soil is lost. Any soil incapable of producing a crop of weeds must also be incapable of sustaining any other form of plant life, and this should be a regular guide and assessor of soil fertility. The advice given in Chapter 7 and 12 will show how best to manage a weed crop to secure the maximum benefit.

Monoculture

Monoculture is another practice abhorrent to nature, whose free admixture of all forms of plant life is a noticeable feature of her methods and a joy to all who may be privileged to behold it. Yet nature is ruthless in upholding her doctrine of the survival of the fittest in order that all species may continue in health and vigour. Her planting is a model of perfection and her choice of plants an example to mankind in general.

Think for a moment of what goes into the formation of a wood: seeds of grasses and a host of lowly plants are sown together with those of the mighty oak. From the outset a grim struggle for existence is the order of the day, and any plants which fail by reason of weakness and ill health (which is but another name for disease) are removed and by their death and decay provide sustenance for the survivors. The process of elimination begins during the early lifetime of the natural wood, where first to disappear are the annual and the shallow-rooting plants which are crowded out by their more robust neighbours. This weeding out of the weaker members is gradual but thorough. For a short time the floor of the wood is bare of all carpeting plants, because the leafy canopy of saplings and trees prevents the rays of the sun from penetrating to the ground below. Slowly, but relentlessly, the process of elimination is carried on. The Ash, Beech, Elm, and Oak encompass the downfall of the shrubs, and their disappearance once more allows the sunlight to reach the floor of the woodland; the cycle of life starts all over again with the growth of grasses, bracken, and so on. At this stage the bluebells and wild garlic make their appearance and the green carpet is re-formed; the bulbous plants quickly overcome all rivals and hold undisputed sway for a time. It would seem that at long last nature is satisfied with her handiwork, but in this changing world no achievement is final and permanent. Death in some form comes to all things and the long-lived oak must sooner or later perish and return to dust. When the tree falls the wild garlic also must pass away for without the kindly shade of the branches overhead it quickly withers. Then we see the reappearance of the grasses, and with them, seedlings of the mighty oak and other trees once again. We shall find no acres of oaks, but instead an ideal mixture of plants of all kinds, and this should provide the perfect example for man to follow instead of the customary segregation of different species.

The Unity of nature

The lesson we should be at pains to learn is that nature bases all her operations on humus, into the formation of which go bits and pieces from a wide range of plants. Natural growths of varying heights give shade when it is needed and afterwards deposit the autumn leaves upon the ground for the nourishment of the soil. At the same time the roots, as well as aerating the subsoil by their deep penetration, draw as if from an inexhaustible well and send to the branches and leaves, in the sap current, the life-giving elements of plant food prepared by mother earth.

Can we not model our system of cultivation more upon these practices of nature by allowing the weeds to act in the same manner in our gardens as do the trees in the wood? We have seen how during the lifetime of the wood, crop follows crop in endless succession, each in turn contributing something to the common fund of soil fertility. Unless man interferes, the circle is complete and the majesty of the wood moves with a perfect rhythm and purposefulness. Awe-inspiring in the intensity of her desire to reach her objective, ruthless in pursuit of her ideals and with a grandeur unmatched by all the science of man, nature unfolds her panorama before our eyes.

How feeble in comparison are the efforts of man! His painstaking weeding, his artificial watering and synthetic feeding are a direct contradiction of the methods adopted by the supreme gardener - "Nature". It is no wonder that failures are so common in a system so divorced from that which mother earth has chosen to follow.

Recognition of the necessity of humus, and the futility of monoculture has enabled me to grow crops of a high standard and, at the same time, maintain the fertility of the soil. I have proved that the companionship of weeds is a helpful factor in the cultivation of the common garden vegetables. Nature may be trusted to remove all surplus plants when they have served their purpose; yet man in his short-sightedness too often destroys the weeds for no other reason than that they are termed weeds by the textbook. For the sake of convention - keeping the garden clean - he is committed to a policy of replacing, by the purchase of synthetic material, the natural elements of plant food he has needlessly destroyed.

This contention may not appear to be very closely related to the practice of gardening proper, but as I pointed out at the beginning of this chapter, nature is one and indivisible. Most of us are too prone to establish differences and distinctions where unity of purpose should prevail. The operations of nature, the waste products of industry and of the countryside, the germination of seeds, the growth of the seedling, the maturity of the plant, its fruition and its decay, are linked together in this endless chain of events. If any one link can be considered of more vital importance than the rest, then I declare that link to be the return of all waste materials to the land.

Chapter 2

SOILS

The condition of the soil will in no small measure determine the value of the crop, and it is to the advantage of the gardener to find out as much as possible about the soil in order to make full use of his land.

Soils are formed by the combination of many ingredients - mineral particles and their decomposition products, decayed vegetable and animal matter, and water. In some gardens the mineral matter is in a fine state of division (sand), in others it is coarser (gravel), while here and there it is in larger pieces still (rock). Perhaps the most important of the products derived from minerals is clay. By far the greater portion of what we call soil, has no value as plant food, because it is insoluble. It is the organic matter derived from animal and vegetable wastes that we mostly rely upon to produce our crops, though water is also essential. In peaty soils we find an abundance of decayed vegetable matter with very little sand, gravel or rock.

Maintenance of Fertility

Constant production of crops cannot be carried out for long without some system of replenishing the plant food that has been absorbed. Therefore we have to devise some way of returning to the soil what the plants extract from it. A healthy, fertile soil contains many elements essential to plant life, but many gardeners only feel concerned about three: nitrogen, phosphate and potash. Besides these elements, and the sand, gravel or rock, the soil is the abode of a vast population of earthworms, fungi and soil bacteria, so it will be seen that no matter what substance we use for the enrichment of the soil, we must not look upon this as a plant food only. Certain substances may supply nitrogen, phosphate, and potash, without feeding this vast soil population, or improving the water-holding capacity: in this case they are less beneficial than a natural fertiliser which, while nourishing the plant, also feeds the population of the soil and makes it more capable of retaining moisture.

Let us give a simple illustration of what is meant by improving the soil. Consider, for instance, a garden where the ground is sandy or gravely. Soil of this character is usually dry because the fine particles of sand cannot hold the rain water, but allow it to drain away too quickly for the plants to avail themselves of it. Apart from the fact that such soil is dry, it is usually lacking in plant food; crops grown on sandy land mature early, but the yield is often small. We can add to this type of land the ordinary commercial fertilisers to help in increasing the weight of crop, but this does not improve the capacity of the soil to hold water, neither does it provide food for the earthworms or the various mico-organisms. It is evident, therefore, that the use of chemical manures is a very one-sided effort in the direction of soil improvement. If instead, farmyard manure is applied, we feed the earthworms, the plants, and also at the same time increase the water-retaining capacity of the sandy soil.

The supply of farmyard manure, however, is often very limited: we are compelled to make a little go a long way, which inevitably slows down the improvement of the land. For this reason the wise gardener will adopt a system of mixing vegetable waste along with farmyard manure in what we call a compost heap. If the composting is carried out correctly, the soil will derive more benefit than it would if only the farmyard manure was used. Animal manure from a farm loses much of its goodness before it reaches the garden, as it is usually piled up outside and exposed to the weather, until a great deal of the more soluble substances are washed out by the rain, dried out by the wind and sun, or evaporated as a result of fermentation.

By making good compost from animal manure and garden refuse on the lines described later in this book we can quickly achieve an improvement in three directions: (1) we can increase the weight of crop; (2) we can provide a natural diet for the soil population and (3) we can improve the moisture-retaining capacity of the soil. A further point in favour of compost making is that it assists in keeping the garden clean and tidy by making full use of the many scraps of waste material which might otherwise be difficult to utilise.

Types of Soil and their Treatment

Compost will improve any and every type of soil, whether its chief characteristic be sand, peat or clay, as long as we understand how to prepare the compost for each type. To improve a sandy soil, heavy dressings of well-ripened compost should be used: this will feed the population of the soil first and afterwards the plants, as well as making the soil more capable of holding moisture during the summer. For peaty soils frequent dressings of quicklime are advisable; this reduces the acidity and releases plant foods which would otherwise be locked up in the humus, and, in fact, such land will continue to grow good crops of vegetables for quite a long time with little in the way of manures but fresh lump lime. Heavy clay soils are naturally rich, but always difficult to work with, and any fertilising material added to them should be of such a character as to open up the texture rather than to supply food. Sand and old mortar rubble are often recommended, but only the fortunate few will be able to obtain sufficient quantities to make any appreciable impression if the garden is fairly large. Burning a portion of the top soil is also beneficial, but here again the amount which the average gardener can burn each year would be too small to influence the whole area. Sufficiently large quantities of straw, together with woody material such as the stalks of herbaceous plants, if dug in each year before they have fully rotted, would do much to improve the land, but even this treatment may be beyond the reach of many. Regular dressings of lime will help, but one thing which should never be used on clay are the ashes from a garden bonfire, as this will only increase the stickiness of heavy soil, or even of light soil if used too freely.

Years ago I worked in a vegetable garden which had been completely ruined by an over generous use of wood ashes. The soil had been a beautiful medium loam - the best of the soils for vegetable production - but in ten years it had been so changed in character that it was harder to work than most clay soils. The little potash to be found in the ashes in no way compensates for the loss of texture, and unless one is experienced in the use of wood ashes it is best to leave them severely alone. They are valuable as a top dressing for black-currant bushes because the mass of fibrous roots which these plants throw out prevents the soil particles from running together, but for vegetable growing on most soils, the gardener with little experience will find wood ashes more dangerous than useful. They are best used in the compost heap.

Too much advice on soil treatment is often offered on the basis of too little practical experience. Advice should only be given when it is founded on sound personal and practical knowledge. It is the person who follows, and not the person who offers unproved advice, who will be the sufferer.

Functions of the Soil

Let us take a common-sense view of the soil: think of its functions; how it must provide an anchorage for the roots of plants; how it must nourish the plants; and remember the teeming millions of its population, all of which must be fed. The older generation of gardeners believed that 'muck is the mother of all vegetables', and their creed is sound. In your treatment of the soil you will not suffer if you forget all you know of artificial fertilisers and chemical concoctions. They can never support the soil population and can be dispensed with in any well-run garden.

Let me give one instance which recently came to my notice. Permanganate of potash is claimed to be a substance having two functions: (1) as a fertiliser, and (2) as a worm killer. But obviously if it fulfils its function as a worm killer, it cannot, by the laws of nature, be of much benefit as a fertiliser; and I, for one, would have little use in the garden for any substance that killed the earthworms.

The Importance of Humus

It is my experience that no lasting benefit can be found in any kind of fertilising agent which does not eventually form humus. If only one aspect of the soil is to be considered, let that aspect be the humus content; beside this all else is of little consequence. The moment we lose humus we lose fertility. Indeed, fertility is humus. A healthy plant is only possible where humus is to be found, and no amount of science can deny this simple fact. The soil demands the quick return of all wastes in order that it may produce maximum crops of maximum quality. If we eat quality crops from healthy soil, we reap the benefit, and in my opinion there is no place in the garden for a single ounce of artificial manure - as it cannot give health to the soil. Let us remember that the plants can do without us, but we cannot do without them.

Profit and not health is the motive which inspires all the propaganda relating to the use of chemical fertilisers. Nature alone reveals the secret of improvement in the soil: she has nothing to sell, but much to teach. There is truth in the old saying, "There's none so blind as those who won't see", and it is the mark of a true gardener to follow nature's lead rather than to accept whatever man has to sell.

Chapter 3

MANURES

The function of manure is to make the soil more productive. This does not mean merely to produce a wealth of vegetation, such as is often the case with stimulating substances like sulphate of ammonia and nitrate of soda, but rather to increase the humus content of the ground. And for this purpose two things are essential - animal manure of some kind and vegetable matter.

Compost

It stands to reason, therefore, that compost, which includes both these ingredients in a form easily digested by the soil, is a manure which can be recommended with the utmost confidence. Compost, as will be seen from the full description given in a later chapter, is an adaptation of nature's method of manuring the field and the forest. Here the wild plants and trees enjoy a freedom from disease and a degree of productivity which, compared with many man-cultivated parts of the earth today, is little short of miraculous.

I would beg those gardeners who are cautious about making compost lest it should prove too complicated and difficult, to put aside their unnecessary caution and try their hand. It is true that there is an art in compost making and the more care that is put into it, the better the result will be. A valuable product can be turned out under all kinds of conditions, and to forgo the undoubted benefit of a humus forming manure because it is not of the first rank, is mere foolishness. A little perseverance through early attempts will soon bring its own reward.

Farmyard Manure

Compost, moreover, has the advantage that it makes the most of what little farmyard manure may be available. Mixed with the vegetable residue of the garden to produce compost, it will go much further. Yet farmyard manure is in itself a more or less complete fertiliser and its value in the garden has always been recognised and appreciated.

To maintain its richness, however, we should take care of our manure heap and not allow it to deteriorate through exposure to wind, sun and rain. Its nitrogen content is easily washed out by the rain, or evaporated by the warmth of the sun, while at the same time, the heap, if unprotected, probably ferments and diminishes in bulk, so that a loss both in quality and quantity ensues. To save such wastage, cover the outer surface completely with soil to a depth of four or five inches: this will check evaporation and retard the process of decomposition. When the manure is needed, it will be found that the heap goes further and is richer for having been covered with the earth.

Commercial Manures

With compost, and farmyard manure, any garden can be maintained in real fertility, but there are also certain organic manures of what we may term the commercial kind, which will give beneficial results for a time. It is my opinion, however, that their use should be sparing and that they should not be expected to take the place of organic substances.

Fish meal is the best of this class and is a balanced or complete manure, but it is not wise to use it in the production of onions or there will be difficulty in ripening the crop. It may be used with advantage in the early spring for cabbages and similar crops to hasten their growth. This organic fertiliser is infinitely preferable to nitrate of soda for it has no deleterious effect on the soil population.

Bones can usually be obtained in the form of bone meal, steamed bone flour, or dissolved bones. Of these I prefer steamed bone flour which is excellent for onions, if applied as a dusting during February the early ripening of medium sized long-keeping bulbs will be greatly assisted. It is also beneficial for all fruit trees and bushes. Bone meal acts as roughage in sticky soils, helping to keep them open and well drained, but it is very slow-acting and should, therefore, be applied to the ground well in advance of crop requirements. Much quicker in action is the form of bone manure known as dissolved bones, or bone. This is powdery substance formed from raw bones treated with sulphuric acid, but it is not to be recommended.

It is largely for their phosphoric acid content (though they also contain a small percentage of nitrogen) that bones owe their popularity as a fertiliser, but I doubt whether most gardens really need more phosphate than a good sample of compost contains. If I thought the addition of bones in some form to be necessary, I would choose steamed bone flour to remedy the deficiency, but after using any form of bone manure for a year or two further applications will generally be found to be a waste of time and money. Observation of the results usually indicates a marked improvement after the first dressing or two and thereafter no noticeable beneficial effect. To apply bones to the garden every year is an unnecessary expense.

Dried blood, hoof- and- horn meal, and *rape meal* may help to supplement the supply of farmyard manure where the land is poor, but the limit of their effectiveness is soon reached and they are only of real service where the humus content of the ground is low. A good dressing of compost would quickly render their use unnecessary.

Inorganic or Artificial Manures

The other commercial manures in general circulation are of the inorganic class. They have become the mainstay of many gardens and allotments - easy to purchase, easy to transport, easy to apply, and to the casual observer usually productive of rapid and remarkable growth. But that is only one side of the picture, and my conviction, after personal experience of such fertilisers, is that there is no place for them in a well-run garden. These artificial manures fall into three main groups: (1) nitrogenous fertilisers, (2) phosphatic fertilisers, and (3) potassic fertilisers.

Nitrate of soda has a forcing effect and for this reason is much used by the exhibitor of flowers and vegetables. Its effect on the plant is quickly apparent, but of short duration. A sprinkling round young spring cabbages will be followed by noticeable developments in three or four days: the foliage will turn a deep green and the plant will enter a period of rapid growth. In this way many market gardeners push on an early crop which finds a ready sale in the spring, after which the ground can be cleared in time for some later vegetable. But in the private garden or allotment this gain in time, which at most is only a few days, can never be worth the risk of pests and diseases, for which nitrate of soda must be held responsible. For there is a debit as well as a credit side to the balance-sheet. Nitrate of soda, with its forcing action, produces a 'proud' plant and "pride goes before a fall".

I cannot find any foundation for the old saying that whatever attacks our plants 'knows what's good': rather the reverse is the case. Wood pigeons seldom attack young plants of cabbage, cauliflower, or sprouts if they are healthy, but seize upon any that are infested with club root or gall weevil, or have received a check from faulty transplanting.

I can give an instance, a   case of cabbages in two neighbouring gardens, one of which was dressed with nitrate of soda and duly responded to the treatment, but later the whole crop was ruined by caterpillars. Those in the other garden, which received no artificial dressing, were unaffected, though both sets of plants had come from the same seed bed.

Sulphate of ammonia is another nitrogenous fertiliser which is widely used and often produces marked and rapid growth of vegetation.   Its effect on the soil, however, is disastrous; it robs the soil of lime, so that it quickly becomes acid and sour unless more lime is repeatedly applied. But some forms of lime burn up the humus in the soil, again impoverishing it and robbing it of the power to confer disease resistance on the crop, so that the latter state of ground is worse than the first. There is a distinct connection between the use of sulphate of ammonia and certain plant pests, and it is a significant fact that where this form of nitrogen is applied club root and gall weevil will inevitably be found.

Superphosphate of lime and *basic slag* are the best known and most widely used of the phosphatic group. I once used considerable quantities of both, yet, although plenty of farmyard manure was being applied with these artificials, in a comparatively short time the soil rebelled and persuaded me to give them up entirely, and though my observations have led me to the conclusion that the superphosphate was more responsible for the bad effects on the ground than the slag, I have since obtained distinctly better results without the use of either.

The first indication of an adverse change in the garden was observed in the earthworms. Not only did they decrease in number, but the varieties changed, and soon it was rare to see a normal, healthy, vigorous worm, capable of its vital work of draining and aerating soil by its burrowing. Then the physical character of the soil rapidly deteriorated. In spite of regular trenching and ridging it became stubborn and harsh, working badly in the spring, caking badly in the summer, and producing less and less. Although a full dressing of lime was given regularly, the weed crop showed significant signs of sourness, daisies appeared; groundsel flowered when about two inches high and the same piece of ground produced three or four crops of groundsel in the season. Where vegetables were planted the cauliflowers buttoned very early, cabbages grew hard and tough, peas were short in growth, and lettuces leathery. Moreover, the garden became infested with slugs to such an extent that every bit of green leaf was eaten; 10,000 onions disappeared in less than a week; potatoes, when lifted, were mere shells and few were fit to store; the slugs even climbed the gooseberry bushes and stripped off the leaves.

During these years labour in the garden was abundant, no routine of good cultivation was neglected, and there were adequate supplies of farmyard manure, so that these disasters cannot but be laid at the door of superphosphate. Nothing but the artificial fertiliser can have been responsible for the death of and soil bacteria with the subsequent destruction of soil texture. I am still patiently retracing my steps, and by building up the fertility of the soil with compost it is gradually returning to a more workable condition.

Muriate of potash, sulphate of potash, kainit, and other potassic manures vary little except in price. The higher the price the less the impurities they contain, but at best, as in the case of muriate and sulphate, the content of impurities is more than 50 per cent and in kainit it may be as much as 88 per cent. Any advantage to be found in the use of these potash salts is far outweighed by the harm caused by the impurities, which are definitely detrimental to plant life, and, in most cases, actually poisonous. Although forms of potash are frequently recommended for onions or tomatoes, I can only describe them as much too dangerous, particularly for amateurs to use.

The general effect of these inorganic manures upon the soil, runs contrary to all principles of good gardening.

Though they may appear on paper, by means of chemical analysis to represent the correct elements of plant food, they are not in the form which nature ordains and their use is an intrusion into nature's cycle. The soil cannot properly assimilate them and their effect is always to burn up the humic cement of the compound soil particle, and destroy the texture of the earth. Even when applied in conjunction with farmyard manure, one wages war against the other, the land is divided against itself and cannot be expected, in the midst of this conflict, to produce a normal, healthy crop.

Chapter 4

THE EFFECTS
OF ARTIFICIAL FERTILISERS

Before considering any adverse effects on our crops, which appear to be due to the application of artificial manures, we must ask what benefits have been derived from their use? Are they time-savers? Decidedly not; for any time saved is lost in repeated spraying to combat diseases brought about by their use. Do they produce heavier yields? Not over a number of years, as compared with crops grown in naturally fertile soils. Do they produce healthy plants? Decidedly not; they are possibly responsible for most of the present day pests and diseases of, for example, tomatoes. Do they improve the flavour of the fruit? On the contrary; tomatoes grown with the help of artificial fertilisers are characterised by their lack of flavour, and I unhesitatingly suggest that no commercial grower of tomatoes would dream of using sulphate of potash on tomato plants if he intended to save seed from those plants for his own use. He knows full well that in a few years' time, if artificial fertilisers of any kind are used, his strain of plants would be of less value.

An Experiment with Tomatoes

A simple test will soon prove my assertions. Grow half a

dozen tomato plants, either in pots or boxes filled with compost, particularly using old tomato stalks and leaves in the heap, and compare them for flavour with any quantity grown with artificials. You will not be disappointed either in weight of crop or flavour. Save your own seed from the half-dozen plants and get even better results the following year. Then try saving seed from plants fed on artificial fertilisers. There will be a marked difference in the third season; in fact, with the seed saved from these plants you will have difficulty in raising a stock after three years. If we compare the time spent in preparing compost for tomatoes with the time saved by using chemical manures and judge the results fairly, we are better able to assess the merits of the two systems. I have tried both and if, by the labour-saving use of artificial fertilisers, I could produce crops equal to my compost-grown vegetables, I would have continued to use them.

An Onion Failure

For a number of years (1926 to 1940) an area of about one-eighth of an acre had been cropped frequently with potatoes. Artificial manures such as superphospate, potash, and nitrate of soda or sulphate of ammonia were regularly used in   the usual proportions.   The potato crop   was never very satisfactory, but as the same mixture was used on other parts of the garden with identical results, I concluded that this was the maximum which the land was capable of producing. For the 1940 season I was requested to increase the crop of onions and agreed to grow half an acre. This meant reducing the area under some other crop and I decided to cut down the amount of potatoes and purchase them from farms when my own stock was exhausted. Early in March I began to prepare a piece of ground where potatoes had been growing during 1939 and which for a number of years had been treated with a combination of artificial and farmyard manure. Onion seed was sown in shallow drills a foot apart, and the number of rows multiplied by their length came to nearly 1,760 yards. The weather was perfect and the ground in good order, and I looked forward confidently to a satisfactory crop.

Having in the past established a reputation as an onion grower, many visits were paid to see the result of this sowing, and the development of 25,000 onions which were planted out near by. I am more than ever convinced of the truth of that old saying, 'never count your chickens before they are hatched' As day followed day without signs of green in the drills in which the onion seed had been sown, I became more than a little apprehensive and wished I had kept the sowing of this quantity of seed rather more secret. Finally I was cheered by the appearance of six rows of fairly strong seedlings and my confidence in the remainder was restored. I ordered rubber rings for fastening the bunches of thinnings which were to go for sale after carefully estimating the required number. I have to admit that most of the rubber bands are still in their boxes, for never a bunch of thinnings saw the market. Out of the forty-odd rows sown, only two or three at one end ever produced any worth while plants. Where artificial manures had been regularly used the crop failed completely, and the only rows which produced plants were those which were just outside this plot where the potatoes had been grown with chemical fertilisers. These were on a piece of ground which had been treated with sewage sludge many years before. Every crop which occupies the old potato plot shows its dislike, although no more dressings of artificials have been applied. An intelligent workman, if sent to cultivate that piece of land, can tell to within a yard where the chemicals have been used in the past. The texture is hard and sullen and the soil cakes badly under the summer sun. During the last four years this plot has received two heavy dressings of farmyard manure, but today (1944) it is incapable of producing either vegetable crops or weeds of a satisfactory character.

A Potato Trial

During the early spring of 1930 a potato trial was conducted in the garden, part of which was on the same piece of ground where the onions afterwards failed so notably. In order that the records should be authentic, the trial was carried out under the supervision of the district superintendent of a firm of artificial manure merchants, who supplied the manures, and to whom I am indebted for the following report shown on the next page. As the average yield of potatoes in England is about six tons per acre, plot No.1 might be considered satisfactory, while the higher yield recorded from plot No.2 favours the grower.

During the 1939 season I decided to carry out a further trial with potatoes, using the same piece of ground and the same mixture of artificial fertilisers, the only difference being that plot No.1 in the original trial became plot No.2 in the second experiment. The good results were not repeated, although cultivation was carried out by the same workman. Since then, merely by avoiding this piece of ground I have obtained an average yield of potatoes of four tons from one-third of an area, which is equivalent to twelve tons per acre. Furthermore, the keeping qualities of the crop are so satisfactory that, although the establishment to be supplied has increased in number from thirty to over forty, I have been able to supply all requirements in potatoes from our own stock.

Results of Potato Demonstraion, 1930

No. of Plot	Size of Plot	Manures (per acre)	Yield (Per acre)
1	1/16 acre	3 cwt. Superphosphate 1 cwt. Steamed bone flour 1 cwt. Sulpate of potash	6 tons 14 cwt.
2	1/16	3 cwt. Superphosphate 1 cwt. Steamed bone flour 1 cwt. Sulphate of potash 1 cwt. Nitrate of soda (when planted) 1 cwt. Nitrate of soda (when haulm appeared)	7 tons 16 cwt.

Disease in Black-currants

But that is not the whole of the story. After the 1939 crop of potatoes had been lifted, the area was planted with black-currant bushes, of which we had a very fine stock. My method of planting black-currants is as follows: the bushes are planted on part of one-half of the kitchen garden, which is about 100 yards long; they are set in rows about five feet apart and the same distance from one another; each year about one-sixth of the bushes are dug up, starting at the top of the garden, and the same number of young bushes are planted at the other end; by this means the stock is kept young and productive. During the autumn of 1939 it happened that the piece of ground to be planted with black-currants was that used for my potato trials, and as our stock of young bushes was rather larger than usual we planted the whole of the trial plot area and a little more. It would be revealing to show the black-currants now to those people who advocate the use of artificial manures. Every single bush planted on the old potato plot is smothered with Big Bud, whereas the bushes which are just out of that area   are practically free from infection. The most amateur of gardeners could at once define the area where artificials were applied in 1930 and 1939 simply by examining the bushes. Onions were the first crop to fail, followed by the black-currants.

Stunted Apple-trees

Another striking example can be quoted. Twenty years ago an apple orchard was formed in another part of the garden. After four years a few of the original trees were taken out as having proved unsuitable. One tree of the original selection of varieties did so well that I decided to purchase another like it to replace a tree that was apparently unsuitable for the district. Later, two more of the satisfactory variety were planted, making four trees in all. At that time I was quite enthusiastic about artificial manures and, to a certain extent, on growing fruit for exhibition, and during the next ten years these trees annually received their quota of chemical fertilisers. That I later discontinued their use is beside the point; what I want to record is the strange behaviour of the three trees which were planted after I had tested the suitability of the variety. The latest planted trees, no matter of what variety, are definitely stunted in growth, and although the first trees made rapid progress, others of the same variety planted later are stunted in comparison. Without the example of the onions and black-currants before my eyes, this might well have escaped my notice, or at least precluded my drawing the inference that the continued use of artificials on the site has prevented the later planted apple-trees from making anything approaching normal growth. I have reason to believe that the piece of ground I chose for the orchard had never before been treated with chemicals and that, whereas the earlier planted trees of all varieties made satisfactory growth, the ground gradually became poisoned and proper development of the later trees became an impossibility. This, I am convinced, is the true explanation and it is my belief that no good will come to any crop if artificial fertilisers are used. That two trees of the variety, *Bramley's Seedling*, planted eight years ago, are only now beginning to show satisfactory development must be taken as an indication of the deleterious effect on the soil of ten years' treatment with artificial manures, for this variety can always be relied upon to make vigorous growth in fertile, healthy soil.

The above are only three instances among many which could be quoted. I have yet to experience any ill effects following the use of organic manures, and the only logical conclusion to be drawn is that the safest fertilising agent is "compost".

Chapter 5

ADVANTAGES OF USING COMPOST

If we make our soil fertile by the use of all waste organic material, so that the earthworms and soil fungi and bacteria are allowed to increase, they will provide for our plants all the elements needed to produce healthy crops of full flavour. If we supply food for the worms, they will repay us by creating humus regularly year by year in their own way and their own time, but it must be clearly understood that mineral or chemical manures can never become food for earthworms.

In the wider sphere of agriculture the usual explanation given for the use of artificial fertilisers is that composting requires too much labour. That excuse does not hold good where a cottager cultivates a garden or allotment. The pitman of Northumberland and Durham with the finest strains of leeks in the world, the weaver of Lancashire with his stocks, and the rhubarb growers of the Leeds district have for generations maintained the high quality of their plants without the use of chemical manures. Can the users of artificials show similar results? I have watched the decline of gardening in many districts and in the case of cottage gardening it can undoubtedly be traced to the use of chemical manures. The standard of vegetable culture is appreciably lower than it was even thirty years ago. The few outstanding exhibits at the big flower shows represent a very small section of gardening proper, and cannot disguise this decline.

When the subject of compost is discussed, sceptics are always ready to discredit the virtues claimed for this material, but I propose to put on record some of the practical results I have proved compost to be capable of producing - results which are definitely fact and not fancy. There may be compost enthusiasts who broadcast results which are not generally capable of being repeated; if this is so, it is to be regretted, as it does no permanent good and may do harm to others who follow the advice given and then fail to achieve similar success.

Quality in Lettuces

Let me quote one instance where compost excelled in the growing of a crop of lettuce. On the 11th April, 1942 a batch of 500 young lettuce were planted in a row of cold frames. The glass was left off the frames, but the plants had the protection of the frame walls. The soil had been enriched with old compost for preceding crops and was in good condition at the time the lettuce seedlings were planted. Two days later more seedlings of the same age and condition were planted out in two borders - one with a top dressing of fresh compost and the other without. In six weeks' time from the date of planting, all the lettuces which had been put out in the frames were cut - a splendid crop. The 250 seedlings planted on the border which had received a dressing of fresh compost matured just as the frame crop was becoming exhausted. But the 250 seedlings which received no compost were several weeks late and only a very poor sample, not one approaching the quality of the other two lots which had been grown with compost.

In order to see whether these good results would be repeated, more frames were prepared for the 1943 season. Compost at the rate of five average barrow-loads to each 6ft.

x 4ft. light was used, about 250 barrow-loads in all. Some frames carried two crops of lettuces and others only one during the year; but all later produced crops of radishes; one row grew marrows, and one cucumbers. Four rows of frames altogether were planted with lettuce, and the crops were quite remarkable. After two crops of lettuce, one row produced five lots of radishes in the year and saleable radishes were being grown in mid-December. These lettuce and radish crops were without exception the best I have ever seen.

The materials used in making the compost were chiefly perennial weeds with a fair proportion of hard, woody material, sawdust, and a very little farmyard manure. Much of the woody matter was old peasticks, white with fungous growth - a material many gardeners would have feared to use - but the soil in those frames today rivals the finest humus any woodland can produce, and the same good results, which I anticipated this year, without any further addition of compost have been obtained. After this year a renewal of compost will be necessary, not because the soil will have lost its fertility, but because the soil level will have dropped too low to be serviceable for another lettuce crop. A heap of waste material will thus produce the finest possible crops for two seasons.

This year's trial (1944) has just been completed. Seedling lettuces of the variety *May King* were planted in April under fifty yards of low, barn cloches, and under the same row of cloches cauliflower seedlings were also set. Three weeks later, on a border dressed with compost, chrysanthemums were put out, and between each plant a lettuce seedling was planted. Although these latter were not protected in any way, they matured at the same time as those under cloches. About the same time also, three rows of cauliflower plants were put out in line with those covered by the cloches. When the cloche-grown plants were ready to be uncovered, their size was no greater than that of the plants which had not been protected by cloches. The ground selected for the crops under cloches had been used for a number of years before the war for growing chrysanthemums and had been regularly stimulated with chemical manures; the other two plots had not had any chemical manures to stimulate production.

Preventing Maggot Attack

During the spring of 1941 a quantity of autumn-sown cauliflowers were planted on a warm border for the purpose of securing an early crop. They developed into strong, fine plants, but 75 per cent were ruined by maggots. A similar crop was tried on this border in the spring of 1943, but before planting the ground received a normal dressing of freshly made compost. The cauliflowers were ready by the first week in June and not one plant failed to mature. Other growers in the district who do not rely on compost again suffered heavily from plants 'grubbing off'.

Cure of Virus Disease in Raspberries

About twelve years ago I planted some raspberry canes of the *Lloyd George* variety, which were stunted and so badly infected with virus disease that the grower was only   too pleased to give them away. They have recently been inspected by a representative of the Ministry of Agriculture and Fisheries and pronounced 100 per cent healthy. But at the time I planted the canes this variety was showing unmistakable signs of degeneration, and growers were awaiting the introduction of a new one. In due course a variety appeared under the name of *Norfolk Giant*, but apparently its run of success is now in question and the time has again arrived when growers are at the cross-roads, or more correctly, at their wits' end to know what to do. Obviously future plantations of *Norfolk Giant* are out of the question, as experience has shown that the first sign of failure in any variety is followed by speedy degeneration.

I chose that moment to plant *Lloyd George* - a moment when it was being discarded in favour of *Norfolk Giant* - and was determined to satisfy myself as to whether the cause of the failure was really due to failing vigour on the part of the variety, or to a wrong system of cultivation.

I am satisfied that *Lloyd George* is constitutionally sound, and that where any evidence appears to support the idea that it is worn-out, the cultivation has been unsound. Judging by the present appearance of my canes, I see no reason why the plantation should not retain its recently won health and vigour for many years to come. I may mention that, as a variety suitable for my part of the country, I do not consider it equal to *Baumforth's Seedling*, because a large percentage of its fruit ripens too late to be of much use. Raspberries in September are possible in Westmorland, but early August is late enough to ensure good berries, and the ideal variety for this part of the country would be one producing ripe fruit early in July and finishing by the end of that month. *Lloyd George* is too prone to give a flush of fruit in early July, a second towards the end of the month, with a later batch trying to ripen in September. In 1942 and 1943 my last picking of sound berries was on 4th August, but each year upwards to 100 lb. of fruit has been spoilt during August and September on account of the weather.

Below I give the yield obtained from 100 yards of canes during the last ten years:

1934	115 lb.	1939	137 lb.
1935	128 lb.	1940	107 lb.
1936	273 lb.	1941	118 lb.
1937	238 lb.	1942	249 lb.
1938	153 lb.	1943	214 lb.

As will be seen above, the year 1936 shows the highest yield, with 1937 slightly lower, and it was not until 1942 that

I passed the 200 lb. mark again. During the autumn of 1936 I commenced alterations in the pleasure grounds of Levens Hall which kept our small staff fully employed, and in consequence the raspberries were neither cultivated nor mulched until the autumn of 1940.

During May 1939 I answered the appeal to rejoin His Majesty's Forces and was discharged on medical grounds a year later. I then decided to become personally responsible for the care and cultivation of the raspberries. In the autumn of 1940 I cut out all the old fruiting canes, tied in the new canes, and cleared the plot of weeds. The rows were mulched with compost and the space between the rows forked up. Early in April another was given, and the space between the rows forked again. A very slight increase in crop resulted. This method has been repeated each year and has resulted in a heavier yield of fruit - much heavier, in fact, than my figures show, as a good proportion of the crop each season could not be gathered on account of bad weather. Rain fell on twenty-nine days during August 1943, which was well above the average even for the Lake District.

I believe my modest test has clearly proved that a return to the use of compost must be made if we are to retain good cropping qualities of our stocks of raspberries over a reasonable period of time. I have been told, however, that I have not succeeded in curing the virus disease, but only in masking the symptoms. My reply to this objection is that I am satisfied that by the use of compost I have secured good crops for the past ten years from canes which, when planted, were incapable of producing fruit. Even if compost does no more than mask the disease, the fruit my canes have produced would do credit to any gardener, and the practical results I have obtained seem to make the difference between masking the symptoms and curing the disease relatively unimportant. My crop record for the past ten years has at least proved that the benefits to be derived from compost are not short-lived, but rather cumulative.

A Fourfold Test

A fourfold test is now being carried out on a border 200 yards long. A number of pear and plum-trees are trained along the wall, and in front are roses and herbaceous phlox. The ground had its first dressing of compost in the autumn of 1942, but in former years the fruit trees had been given artificial manure. All went well for a time, but later the pears developed scab, and the plums were smothered with greenfly every season. The roses were poor, and the phlox ruined by eelworm. No special compost has been prepared for this border, but the contents of any bin which was ripe at the right time for application has been used.

Some varieties of the pears, notably *Pitmaston Duchess* and *Doyenne du Comice* showed an improvement after the first application of compost, but *Beurré Six* bore nothing but cracked fruit and scabby foliage as usual. *Marguerite Marillat* produced more growth in one year than in the ten previous seasons, and a cordon of the variety *President Roosevelt* yielded more fruit in 1943 than it has done in all the years since its planting in 1919. After two dressings of sawdust compost - I call it sawdust compost because the mixture contained 50 per cent of this material - I anticipate an all round improvement in 1944.

As far as the plums are concerned, they were not smothered with greenfly, but their cleanliness might be accounted for by weather conditions and not the compost treatment. Nor did I observe any difference in the roses, though my salesman complimented me on their outstanding keeping qualities. The biggest improvement was in the clumps of phlox. They had been undisturbed for years, the heads of blooms had become very small and the foliage thin and weedy-looking. One dressing of compost unquestionably had an effect; the blooms were better in every way and there was less evidence of the distorted foliage so common in plants attacked by eelworm.

Chapter 6

COMPOST AND ITS PREPARATION

The operations of nature - as explained earlier - are based on humus.   We can see how humus is made in any piece of mixed woodland. Here the waste products of the vegetation - leaves, bud scales, twigs, flowers, fragments of bark and so forth - form a loose layer on the ground and become mixed with the residues of the large animal population found in every forest. This litter does not accumulate beyond a certain point because it is fermented by moulds and microbes.   Much of it disappears in the process, but an undecomposed residue is left behind which amalgamates with the dead bodies of the fermenting organisms to form humus, the food of the trees and undergrowth, and indeed of all plants.

The wood manures itself.   Nature never provides artificial manures to stimulate growth or poison sprays to destroy insect and fungal diseases. If we copy nature in our gardening we can make our garden manure itself and also produce crops which are not only resistant to pests, but also transmit disease resistance and health to ourselves. This, in brief, is the basis of organic gardening. The foundation of a successful garden or allotment is, therefore, a regular supply of humus.

How does humus act? Its chief function is to feed the

unseen life of the soil - chiefly moulds and microbes - and also the burrowing animals like earthworms, so important in soil. The wastes of this soil population, including the worms, then feed the plant. They also maintain the tilth or texture of the soil. This is accomplished by the dead bodies of the microbes - minute grains of glue-like substances - which cement the fine particles of soil together to form large compound particles. In this way the structure of the soil is maintained. Without this cement the soil structure would collapse and we would have a dense soil instead of a free-working one. But it is not sufficient just to feed the soil population and to maintain the tilth. The soil needs a reserve of additional humus to absorb and retain water and to keep the ground warm.

How can we copy the operations of nature as seen in the wood and prepare sufficient humus for our garden? By composting all the vegetable and animal residues on which we can lay our hands. The humus in the resulting compost, when added to the soil, gets used up, so it is essential to have some easy continuous process of making it, which at the same time helps to keep the garden or allotment neat and tidy.

The ideal method of composting for a small garden or allotment is to use a couple of bins side by side; the purpose of the second bin being for ripening the compost.

The reasons why composting in small gardens is best done in a bin are these. Firstly, because the supply of waste material is small and intermittent. The size, of an open compost heap would therefore be small, and the ratio of surface to volume will be high. A heap, being open to the elements, will therefore, be kept too cold by wind and rain in spite of the fermentation going on inside. Secondly, excessive rain will keep the heap sodden and replace the air in the various wastes by water. These two problems - constant cooling of the mass and the interference with the air supply needed by the moulds and microbes - make it necessary to keep the small compost heap warm and also to regulate the water supply. All this can be done in a suitable bin or box. These precautions are not necessary in large heaps - as a large heap protects itself.

Two suitable boxes or bins can be made as follows. Both are exactly the same size, so the following description applies to both.

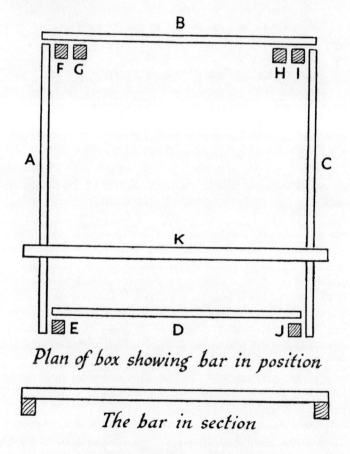

Plan of box showing bar in position

The bar in section

The New Zealand Compost Box

Making the Box

Materials required: Six 3 ft. 3 in. lengths of 2 in. x 2 in. for uprights. Twenty-four 4ft lengths of 6 in. x 1 in. board for the four sides of the box.

A, B, and C are the sides, each consisting of six boards, each 4 ft. x 6 in. x 1 in., nailed to the uprights half an inch apart to allow ventilation.

D is the loose front (six boards)

E, F, G, H, I and J are the uprights (each 3 ft. 3 in. long).

K is the bar, provided with a block at each end, to sit on top of the sides A and C to stop them spreading.

The unplaned timber should be oiled with old motor oil to preserve it, but tar or creosote should not be used.

The box, which has no bottom, stands on the ground. First nail the side **A** to the uprights **E** and **F**. Next nail the back **B** to the uprights **G** and **H**. Next nail the side **C** to the uprights **I** and **J**. When nailing the boards on the uprights leave a half-inch gap to provide ventilation. The three sides of the box are now complete. The sides and end are bolted together by means of four bolts - each fitted with two washers and a nut which unscrews on the outside - which join the back **B** to the uprights **F** and **I**. The front **D** is made up of loose boards 6 in. x 1 in. slipped behind the uprights **E** and **J** as the heap rises. To prevent the sides **A** and **C** from spreading outwards use a wooden bar 2 in. x 1.5 in. with two wooden blocks (3 in. x 2 in. x 1.5 in.) as indicated in the diagram.

If the box has to be moved to a new site, remove the loose boards and the four bolts and re-erect the box in a fresh place.

How much compost can be made in a year in a pair of these compost bins? At least four tons. We need never weigh compost as it can be more easily measured by volume. As a general rule two cubic yards (54 cubic feet) of compost weigh one ton. For a medium-sized garden a pair of two-ton bins can be made out of old railway sleepers. Ideal measurements would be 6 ft. x 6 ft. and are 3 ft. 3 in. high.

Choosing the Site

The choice of site for a compost heap is a most important matter and should only be settled after ample consideration, so that the best possible position may be selected. Naturally a central site has many advantages and is always to be recommended. If the selected place is at the end of the garden, this may involve a considerable distance to carry material for the heap and also to convey the finished compost when it is applied to the ground. A place near the house is convenient, but may be objectionable to some people. All these points should be taken into consideration by the gardener concerned, who alone can make the best decision in a particular case.

Wherever the box is placed, the base must be dry: this is an important point. Should the only available site be a low-lying part of the garden, the base should be raised well above the surrounding ground before erecting the box. It may be necessary to wheel a few barrow-loads of broken stones or rough cinders to form a high and dry base. In districts where bricks are available nothing could be better, but it is essential not to place the bricks too close together. Air will then be admitted to the heap which is desirable and also the contents of the box can easily be cleaned out from such a level base. If a good base is made in the first place, future operations will be simplified.

Where it is impossible to get sufficient wood to make a whole box or bin a site next to a convenient wall could be selected: the wall then acts as one side, or the angle of the wall will take the place of one end and one side.

If possible, a level site should be chosen, as sloping

ground in showery weather can be dangerous. The task of filling and emptying the box is easier when the surrounding ground is level.

Material Suitable for Composting

It may be of some help to beginners in compost making if I indicate the various material that I have used successfully. Lawn mowings, all kinds of rough grass, fallen leaves, all kinds of annual weeds, and damaged or surplus vegetables form the soft green matter. The stalks of vegetables, such as cabbage, and the softer prunings like young raspberry canes, the summer prunings of apples, pears, etc., as well as those from soft fruits; gooseberries, for example; may be used. In the autumn the stalks of herbaceous flowering plants are particularly valuable if cut down while green with the sap still running, and all flowers which have been used in vases in the house. Kitchen waste, such as tea leaves, coffee grounds, eggshells, etc., sawdust from any kind of wood, and damaged straw are all useful ingredients. The winter prunings of fruit trees, clippings of box, ivy, laurel, yew and other evergreens, as well as roses and climbing plants, may be incorporated in the heap. The waste from greenhouse crops, such as the stems and leaves of tomatoes and cucumbers can be included, and I have also used old black-currant bushes which had been stacked in a near-by wood for about a year. Raw peat may supply bulk, also wood ashes, and seaweed may be used when available.

Straw is a very suitable material for mixing with green vegetable waste, but too much in one box is not to be recommended. It does not decompose quickly, owing to the fact that the stems are protected, when growing, by a kind of natural varnish which must be destroyed before the straw will absorb water and rot, but otherwise it is first-rate material.

In my own compost bins I use many tons of sawdust, but only after it has been weathered. To do this, it is best spread out in a yard about three inches deep, or around the compost bins where it absorbs any surplus liquid from the bins and becomes mixed with the earth, discolouring and rotting all the time. The moment when it is ready to go into the heap is indicated by the presence of worms. When worms are found in the sawdust, one may be sure it is sufficiently weathered.

Compost made from straw, sawdust and a little manure is particularly good for light soils and will give lasting results. I know of no better mixture for producing quality potatoes. The skins of which, when lifted, are as clean as if they had been washed, and perfect tubers of exhibition quality will be formed.

Although I have discussed the usefulness of sawdust in compost heaps, I would emphasize the fact that only small quantities of weathered sawdust can safely be incorporated in compost for poor land, or ground where compost has not previously been applied. The safest crops to grow after a first appilcation of compost containing sawdust are peas or beans; but as a mulch this type of compost may be used on any crops. I have used uncomposted, weathered sawdust as a mulch for spring cabbages and blackcurrants with excellent results. I have ploughed in raw sawdust and planted various crops, but the only vegetables that did not resent this treatment were peas and broad beans. If sawdust is used intelligently, it is excellent for the ground. It will be seen, therefore, that I have made use of everything a normal garden can provide in the way of waste.

As regards manure, anything the country district can supply is suitable - cow, horse, and pig manure, poultry droppings - or sewage sludge can be used. Where animal manure or soiled animal bedding is not available, substitutes such as dried blood, hoof-and-horn meal, or fish manure should be used.

My compost making is carried on in three bins side by side, each about 40 ft. long and 7 ft. wide. My bins, no doubt, are larger than many gardeners will use, so I do not suggest that everyone can compost all these ingredients together; a small bin will be able to deal with a selection from the above list rather than all the items. In the case of a small bin, a little common sense will suggest what ingredients are likely to be most suitable.

I would advise all compost makers to make good use of woody material. If it is not broken down in the first heap, it can be transferred to the next until it has thoroughly disintegrated. To burn such material for the sake of the handful of ashes it makes is wasteful. Woody material has considerable value in compost making in that it prevents the heap collapsing too rapidly, and admits air as well as breeding fungus, which is beneficial both in breaking down the other material and in its plant-food value. This fungous growth is shunned and feared by many, but I have long since proved this fear to be without foundation, for I have grown the choicest plants in a mixture of leaf mould white with fungus. I cannot agree with many of the modern writers who advise gardeners to pick out the twigs and woody material from their wastes, burn it, and add only the ashes. I have for years made a point of incorporating such material in my compost and have never observed any ill effects; in fact I would say that no compost can be considered complete without a certain amount of woody matter.

In making compost we follow the methods of nature and the finished product should always resemble the top inch or two of a woodland floor. What happens to all the twigs and branches which fall to the ground every year in the wood? All is left to a natural progression and is mixed with the other litter on the floor of the wood. This eventually becomes the natural compost we call leaf mould. The law of the woods is that nothing shall be wasted.

A good compost, as discussed earlier, should consist of bits and pieces of everything that one can possibly get hold of, from soft lawn mowings to hard material. The soft matter will quickly become available as plant food and will supply all the food the young plant requires. The hard substances will decay much more slowly, but will assist in soil aeration and supply elements of plant food which are necessary during the maturing of the crop we are cultivating. Between the very soft and the very hard ingredients will be found the bulk of the material used in making the original heap, and this will be called upon to nourish the growing plant over the longest period of its growth. This reveals the importance of a complete mixture. If the bulk of the heap consists of soft or semi-hard material, it cannot, because of its nature or immaturity, continue to feed the plant, and it is therefore essential to make sure that the compost contains all the elements necessary to bring the plants we grow to healthy fruition. Where failures have been recorded in growing a crop by means of compost, this is nearly always found to be the primary cause.

Making the Compost Heap

Having selected the best possible site and set up your bins, the next part of the process is to make a collection of suitable material.

A portion of the roughest material should be placed at the base of the heap to act as a foundation, rather as a farmer does when starting to build a haystack. This will allow a current of air to pass up through the heap from the bottom to the top during the early stages of decomposition, and applies to all heaps irrespective of size. In my large bins I use old tins or buckets to assist in supplying air at the base of the heap, together with a quantity of twiggy material to a depth of about six inches. After that the ingredients should be well mixed. Avoid, for instance, a large amount of lawn mowings in a thick layer, as this will become a slimy mass preventing air from passing freely through the heap. Mix a little of everything you have collected into a small pile on the ground first and then transfer this to your bin or box. Sprinkle a small quantity of old lime into the mixture during the building process, (instead of using a layer of lime to every foot of mixed waste) as this neutralises the excessive acidity. Alternatively, a little earth may be used, but too much hinders the ventilation of the mass.

Where sufficient animal manure is available, use three or four parts by volume of vegetable matter to one of manure. This will make an excellent compost suitable for the production of any type of crop. If poultry manure is used in a fresh condition after breaking up, only half the volume of manure as recommended above will be sufficient. If such substances as dried blood, fish meal, bone meal, or hoof-and-horn meal are used, a double handful to every barrow-load of wastes will be ample. The exact proportion is one per cent of the dry weight of the mixed vegetable wastes. I have never used any of the advertised activators for compost, as I am satisfied that compost of first-rate quality can be made without them. Certainly I would never willingly use sulphate of ammonia, on account of its bad after-effects. After all, artificials have done untold mischief in the soil. Why should we use them in our compost heaps and make artificial humus?

If the compost materials are very dry they must be moistened using a watering can with a fine rose until a condition like that of a pressed-out sponge is reached. If, however, about half of the vegetable wastes consist of fairly fresh, green material, no extra watering will be needed. If a larger proportion than this is fresh succulent green material, it should be withered first and then wetted before use, otherwise silage and not compost will result. A little experience will soon show how the moisture factor in composting should be managed.

The heap should be built as high as the available material permits, particularly if it is an open one and the quicker it reaches its full height the better it will be. Even a small heap, unless it has the protection afforded by the New

Zealand box, should not be less than four feet high: decomposition will be more uniform and the compost will be made more quickly and be of better quality. When the desired height is reached, ventilation holes should be made with a fairly strong stake, pushing it in from the top until it reaches the base, then widening the shaft at the top by working the stake round and round while it is still in the hole. The ventilation hole provides an escape for the carbonic acid gas and allows the intake of oxygen which is required in abundance during the first few weeks in order to set up the right kind of fermentation. If carefully built, a small heap will need only one ventilation hole, but for a large heap it is advisable to make one at three-foot intervals along its entire length.

A small heap needs protection from rain and sun, and if the New Zealand box is used, two pieces of old corrugated sheeting, each 58 in. x 26 in., make an excellent cover. They can be kept in position by means of bricks or stones.

Turning the Heap

The heap will be ready for turning four to six weeks after completion, when it will have lost about one foot of its original height. It should be hot in the centre, warm around the edges, and cold on the top. Make sure in turning that the cold top portion and the warm edges are placed in the middle of the heap, piling the hot part around the edges and on top. The heap may now be left to ripen for another four to six weeks, and this time no ventilation hole is required.

If sufficient material has by then been collected to make another heap, the box can be emptied of its first charge, leaving the turned material to ripen in the second box or in a heap outside while the first box is filled again with the new charge. If outside, the turned heap should be covered with sacking or some material which will protect it from heavy rain and also assist in keeping the heap warm. In the case of the first attempt at compost making, however, I should recommend putting the material back into one of the boxes for three more weeks before it is placed outside and covered, because when a box is filled for the first time it will take a little longer to reach the ripe stage than in subsequent fillings. This explains the advantage of using the same site each time and the importance of selecting the best position for the box in the first place. If all goes well, the compost should be ready for application to the garden in about twelve weeks from the assembly of the heap

The beginner, however, will have to be on the watch after filling his bin for two things: (1) an unpleasant smell or flies attempting to breed in the heap. This ought not to happen and is generally caused by overwatering or want of attention to the details of making the heap. If it occurs, the bin should be emptied and refilled at once. (2) Fermentation may slow down for want of moisture, when the heap should be watered. Experience will teach how much water should be added when making the heap.

Enthusiastic compost makers, especially with their first heap, are usually impatient to see what results it will achieve, but do not expect too much from the first effort. Be careful to notice how your compost looks when ripe and how you can improve it in the next attempt. Follow the simple rules, observe the result, and learn by any mistakes - which are likely to occur with all first efforts. If a gardener is not critical of his own work, he can hardly expect his friends   to indicate improvements for him. When practice brings proficiency, a full composting programme can be undertaken. Soon the inevitable conclusion will be reached - nothing need be wasted and a well-run compost heap can absorb everything that is of no other use.

How to Use Compost

If your compost has been made chiefly from green vegetable

waste without much manure, it will prove most valuable as a mulch or top dressing for a growing crop. Rows of peas, all kinds of beans, cabbages, cauliflowers, carrots, beetroots, are just a few of the ordinary vegetables which will derive benefit from it during spells of hot, dry weather, and it is worth far more than the many gallons of water some gardeners carry to their plants. Most beginners in compost making are anxious to put their first compost to the test and if this advice is followed, they will not be disappointed with the results. Most people make the mistake of using compost too sparingly and try to make a little go a long way. It is much better to concentrate on making one part of the garden rich by using compost freely than to dribble a little all over the ground. In order to get the best results from a garden, a plan is necessary, and the use of compost should be part of that plan. It must be used intelligently if maximum success is to be attained.

Where animal manure is available at the rate of one part to three or four parts of vegetable waste, the result is a compost that will produce good vegetables of any kind, and can be used on the garden at the same rate as old-fashioned farmyard manure. Whether the compost at our disposal is little or much, I do not advocate digging it deeply into the ground. The best results I have obtained by its use have been on plots where it was kept reasonably near the surface. To bury compost deep is to preserve it, for few plants can derive benefit from compost placed far down in the soil and a light loss of goodness goes on continuously. If the autumn or early winter supplies of compost are dug in with the weeds, the weed crop will act as a useful green manure.

The best vegetable I can suggest on which a novice can try out his compost, is a salad crop and lettuces in particular. An easily made compost consisting of a soft material with the addition of a little old weathered sawdust, and a small quantity of animal manure or poultry droppings will produce perfect lettuces, which will mature in record time. Such vegetables as peas, beans, and celery do well if a trench is opened out about a foot deep and then filled with compost. In the following year root crops may be grown on the plot without any further manuring. Potatoes will be particularly successful following any trench crop.

Plan your garden at least two years ahead and so make full use of the material you have available for producing the crops. In considering the way to use your compost, try to concentrate it as much as possible. Get rich quickly should be your aim, but let the getting rich apply to your soil. The selection of crops must be left to the individual, but a system of soil enrichment as outlined above should be followed by all gardeners. Remember that the more compost you can apply, the fewer will be the artificial manures or sprays you are likely to purchase and the better the quality of your crops, for no crop should require the protection of sprays and chemicals where compost is freely used.

I know of no better fertiliser than compost for tomatoes, and have never yet found 'green backs' in this fruit where compost was applied. Cucumbers revel in it: my plants have, over a number of years, averaged one hundredweight of fruit per plant. Chrysanthemums planted out in compost have borne perfect blooms at the end of January. I have observed the effect of compost on all kinds of plants, and only vegetable marrows have shown any preference for farmyard manure when planted out on the flat, that is, by merely taking out a hole and filling with either compost or farmyard manure. It is my practice to make up a rough frame during January filling it to the top with compost; I then grow two crops of radishes and afterwards plant up with marrows. In the autumn the frame is emptied, the compost applied to the garden and the process repeated. This old compost is ideal for planting out a batch of lettuce seedlings under cloches during January, and is very much like old mushroom-bed material in appearance.

As a general rule, however, compost should be applied

to the garden as soon as it has ripened. Avoid storing it if possible but if it must be stored keep it in an open shed and turn the heap from time to time.

Chapter 7

WEEDS

I suppose it is only natural to think of weeds when we think of a garden and also when we consider the making of compost, as without weeds few compost heaps would be possible. There is some difference in the usefulness of weeds for this purpose and we should therefore discriminate between the best and those which are of less value, otherwise zeal may overrun discretion in our efforts to fill the compost bin, with a risk of disappointment when the resultant compost is applied to the soil. By all means make full use of the weeds in compost making, but there are other valuable functions which weeds can fulfil. Let us realise that the weeds can be very much our friends, and need not be the tiresome little growths that cause unending labour which they are usually considered to be.

An Exploded Fallacy

I have been surprised at finding so many uses for weeds, for as a child I was taught to believe that the growth of weeds was possibly the greatest punishment laid upon Adam and many gardeners still look upon them in this light. Everything in nature has a definite place and it is our duty, as gardeners, to find a much better use for the weeds in future than we have done in the past. Frequently I am amused at the amount of sympathy I receive from visitors when they see my weed crops. It is difficult to convince them that I deliberately encourage such growth on any piece of ground not immediately required for food production, for without a covering of weeds on what would otherwise be vacant land much of the goodness of the plot would be lost. The weeds should only find their way to the compost heap when by their size they threaten to damage food crops, or just before they run to seed. On otherwise unoccupied ground they have many duties to perform, all of which are beneficial.

The appearance of strong, healthy weeds in the garden is a natural indication of fertility. The living, healthy soil is a factory, populated by millions of willing though unpaid workers. The natural function of a fertile soil is to prepare food for the crops. If there are no vegetable plants to take advantage of the labour of the unpaid workers, then allow the weeds to develop. Unless plant life is present in some form to absorb the food materials manufactured by worms, moulds, bacteria, and so forth, much will be lost: it is like a factory whose machinery is running without producing anything. We cannot dictate to the labour force in the soil their hours of work and times of inactivity. As long as there is raw material like humus in the soil they can be expected to be at work converting it from insoluble to soluble plant food. By all means let the vegetable plants have first claim upon it, but give to the weeds a chance to store up in their tissues all that the vegetables do not require. This is real garden economy.

Weeds as Sub-soilers

Let us examine another duty performed by weeds - by coltsfoot, for instance. What better sub-soiler can be found in a garden? Think for a moment where you are likely to find this plant growing: in the very stiffest clay it is at its best. Is this not an instance of the wisdom of nature, which causes a plant with long roots to flourish and be at home in just the type of soil which we find most difficult to work? There is no vacuum in soils; therefore oxygen will be admitted at the side of the roots penetrating deep into the very type of soil most likely to require it, and far deeper than we should be willing to trench. The roots, thrusting to a depth of three or four feet, will be instrumental in removing the surplus rainfall, again carrying a supply of oxygen. Nature assigns this task to such plants as the coltsfoot, and as long as we are content to control and not to destroy its activities, it will be beneficial to the garden.

For many years I was concerned with the attempted eradication of a bad patch of this weed in a plantation of raspberry canes. I tried by every means to bring about its destruction, but that my object was not achieved ultimately became a source of thankfulness. In attempting to get rid of the coltsfoot roots considerable damage was done to the roots of the raspberry canes, but the chances of eradicating the weed were slender. I decided to try the opposite process, letting the roots continue their sub-soiling activities while I evolved a method of control. In the spring and again in the autumn, I pulled up as much of the root as I could and all the top, this provided many barrow loads of valuable material for the compost bins. In this way the formation of flowers was prevented, so the question of seed never arose and the sub-soil benefited at no cost to myself and without harm to the raspberry canes.

Another weed which can be turned to good account is the perennial thistle. Like the coltsfoot it is a good sub-soiler and provided that it is pulled up three of four times in the year, it will seldom cause any damage to vegetable crops. The tops of the thistle should be allowed to wither before they are added to a small compost heap.

The annual sow thistle is another good weed for the sub-soil, which will also provide plenty of material for the compost heap. It should be pulled up just before it is ready to flower and then go straight into the bin.

The common white clover, so attractive to the honey bee, also has its value. Though it may not provide as much bulk as other weeds, it certainly will increase the nitrogen content of the soil. The bacteria which invade the roots of this plant are capable of absorbing free nitrogen from the air, so that if only the tops of the plants are cut off, leaving the roots to decay in the ground, the soil will receive a free dressing of combined nitrogen by a natural process, and an extra benefit in the form of honey may well be obtained by placing a hive of bees near the clover.

Yet another good sub-soiler, the dandelion, is perhaps even more valuable. If a large root of this plant is dug up and the soil which has been around it is examined, it will be found to be impregnated with a fungus-like growth. This is a very valuable plant food and wherever the dandelion is found the soil will be enriched. The leaves of the dandelion add a relish to the salad bowl, and the roots after weathering can go into the compost box.

If man did not interfere by destroying weeds he would see that, as in the wood, so in the garden, one particular species would hold sway only for a time. The grasses, the chickweed, and the lowliest growing plants would be smothered by the more robust perennials, which in time would bring about their downfall. The coltsfoot, by the deep penetration of its roots, would aerate the soil until such time as it would have to extend laterally in order to find the oxygen-free soil so beloved of this plant, or else perish. All who have observed the habits of mint will know how difficult it is to induce this herb to remain long in one place, how after the roots of mint have occupied a plot of ground for a number of years, the finest sprigs will be found growing around the outside of the original bed which, in turn, becomes the home of some other plant often described as a weed. The behaviour of any naturally sown crop refutes the assertion that the mint has brought about soil sterility, for often the finest weeds will grow where the mint found insufficient nourishment.

The Advantages of Annual Weeds

Many annual weeds are beneficial and seldom do harm except amongst seedling vegetable plants. They supply bulk for the compost bins throughout the summer months and play a vital part in the economy of the garden.

In growing some crops, such as onions planted out in April, I use weeds and vegetables together, and we have much to learn in this direction. I have found that by far the most profitable way of securing a crop of sound onions is to allow all weeds to develop after about the first week in July. They will compete with the onions for the nitrogen in the soil and assist greatly in the ripening process. Should any weeds show signs of seeding, they can be pulled up and put on the compost heap. The onions will not suffer if they are completely covered with weeds. I can often obtain a barrowful of weed to a barrowful of onions during September, and in fact, the only way to find the onions is to pull up the weeds first; the percentage of onions found with growing tops is very small indeed. I have tried clean cultivation and never found the result superior or even equal to growing a weed crop. The growing weeds, by denying to the onions a supply of nitrogen, improve their keeping qualities, and by digging in the weeds in the autumn a supply of humus is available for the next onion crop in the early days of its growth, just when it is most needed.

Lastly there is another vital duty which a crop of annual weeds can successfully perform. Weeds will prevent the loss of valuable plant food at the periods: (1) while the crops are ripening and do not need much nourishment, and (2) in the late summer and autumn after the crop is removed. At both of these periods the soil organisms are still busy converting humus into soluble food materials. If active roots are not there to take up these substances they will be lost in the drainage, or destroyed by certain groups of soil organisms particularly during spells of wet weather. A weed crop is the very thing to prevent these losses. The food materials prepared by the soil organisms are greedily taken up by the weeds and transformed into succulent and easily decomposable vegetable matter. The combined nitrogen, phosphate, and the potash are safely immobilised in the organic form. When the weed crop in due course is dug in during the early autumn and winter (when the activities of the soil population have markedly slowed down) the soil is enriched with a green crop which, during the early spring, is rapidly reconverted by the soil population into food materials at the very moment the next year's crop needs them. There is in this way no loss - only a temporary immobilisation of the things the crop needs. The weed crop, therefore, acts like a banker, who looks after our surplus money for us and lets us have it again when needed. If we destroyed these late summer and autumn weeds, we should lose nearly half the good conferred on our gardens by our composting programme.

This is not, as some would have us believe, just a pleasant theory; it is a fact which I have proved to my satisfaction by the cultivation of many types of weeds over a number of years. It may be that I have been fortunate in being denied the use of ample supplies of farmyard manure, because I have repeatedly had the task of bringing derelict land back into cultivation with little or no assistance other than that of the weed crops. But if common-sense methods are adopted, this is quickly accomplished. Provided that land of this type is not submitted to any severe strain during the first two years, but is encouraged to grow weeds, its productiveness will be assured.

Times without number the weed crop in the garden under my care has had to be mown with a scythe before the cultivation of a food crop could be undertaken. After mowings had been raked off, the roots of the weeds were turned in without manure of any kind, yet a good vegetable crop was always secured.

It is my rule never to deprive the soil of weeds for longer
than is absolutely necessary for the production of food crops,
and I claim many advantages for this system . Fallow land is
a wasting asset, allowing loss of plant food to take place
rapidly and permitting the washing out of soluble elements by
natural drainage; but the presence of roots of any description
will prevent this wastage. Weeds and compost are
inseparable and together they lay the foundation of true soil
fertility. Elements which have once proved capable of
supporting plant life can be relied upon to nourish plants
again and again, and these elements are to be found in the
tissues of the lowliest weed. The right cultivation of weeds,
therefore, will do much to promote soil fertility which in turn
makes possible healthy crops for human consumption at a cost
which is well within the means of the humblest tiller of the
soil. Raise the quality of the weed crop in a garden and
quality in the vegetable crop will be a foregone conclusion,
for the two are interdependent. It is time that the views of
most gardeners were altered and that a more intelligent use of
weeds was made.

Raising the quality of the weed crop is best undertaken in
this manner. Seed saved from the best plants should be
sown on land which is carrying a poor weed crop. The farmer
is beginning to see the wisdom of ploughing up and re-
seeding old, worn-out pastures. The gardener can safely
follow his example. On poor soil shepherd's purse will reach
a height of only a few inches, diminishing in size as it
increases in quantity. Its tap root is incapable of sub-soiling
the land and the thin tissues of its growth cannot store plant
food elements: such soil, therefore, becomes progressively
poorer. If seed of the common groundsel, taken from healthy
plants, is sown in the autumn, an abundance of material is
provided for the compost heap in the following year and,
although only shallow rooting, it will often penetrate deeper
than poor specimens of shepherd's purse, even though the
latter has tap roots. I have never found controlled weeds

to interfere unduly with the crop sown or planted for human consumption. During the course of the growing season there is room for both crop and weed, if we are prepared to regulate the growth of the latter. Soil enriched by decayed vegetable matter derived from weed crops is equal, if not superior, to soil fertilised by that original substitute for compost - farmyard manure.

Chapter 8

WORMS

To many people an earthworm is - well, just a worm; but to others it is an infallible soil-assessor. I do not claim to be an authority on earthworms, but I have a particular interest in the red, flat-tailed lobworm, and I look for it whenever any patch of ground is being dug over during gardening operations. If, while digging, I can see one lobworm every time a spadeful of soil is turned over, I feel a reasonable satisfaction about that piece of ground. Its presence proves to me that the humus content of the soil is adequate, but its absence gives me good cause to be apprehensive of the welfare of any crop to be sown there. It may be possible to find gold in soil devoid of humus but you will never find a lobworm, and gold to a plant would be a poor substitute for humus. Chemical analysis might perhaps show the soil to be rich in every element necessary for plant growth, but even that would not convince me, for without the lobworm I could not but doubt whether any soil was capable, under ordinary cultivation, of producing a worth-while crop, and I would, without hesitation, describe it as unhealthy ground. The lobworm simply cannot exist without humus in some form: take notice how it multiplies in a compost heap. When turning a heap I have seen worms so numerous that there was practically nothing but young worms in a forkful of ripening compost.

Soil Drainage

In the vegetable garden earthworms can never be too plentiful and the amount of work they perform is out of all proportion to the credit generally given to them. Without earthworms much of our soil would become a sodden, waterlogged mass, requiring the expenditure of much money and time to provide adequate drainage; even then the result would not be equal to that which the earthworms can achieve at so little cost. All earthworms can be expected to play their part in soil drainage, but the best of all is the lobworm. A well-grown specimen is nearly as thick as a man's little finger and is capable of penetrating to a depth of fully four feet. Think for a moment of the immense benefit the soil derives from these miniature tunnels, which conduct surplus surface water deep into the ground, also admitting life-giving oxygen to these lower reaches of the sub-soil and releasing the used-up air. If we derived no other benefit than this, it would be sufficient to make us eternally indebted to the humble earthworm, but that is not all.

Worm Casts

Like other creatures the earthworm must eat to live and its casts, which so commonly disfigure our lawns, are of the utmost importance and extreme value in the vegetable garden. Charles Darwin attributed a vast amount of good to the earthworm. Sir Albert Howard has drawn attention to the fact that the roots of plants, coming into contact with worm casts during their search for food, will pause in their downward thrust to throw out quantities of fine, fibrous rootlets which envelop and penetrate the casts to extract from them the last particle of their goodness. Not all worm casts are deposited on the surface; some are left behind in the old tunnels made by the worm and it is on these that the roots feed. Recent investigations in the United States have shown that earthworm casts contain five times as much nitrogen, seven times as much phosphate, and eleven times as much potash as the upper six inches of soil.

Some Habits of Worms

Worms feed by night and, though capable of moving a considerable distance, they rarely travel far, preferring to feed as far as possible without drawing their tails from the tunnels they have made. Anglers in search of worms for bait find 'body snatching' (as it is often called) very profitable. A torch is useful, for the speed with which a worm can slip out of sight has to be seen to be believed. This the worms try to do at once if disturbed during feeding. I hope I have not suggested to any devotees of Izaak Walton a way, otherwise unknown, of securing a bag of worms for their next fishing expedition, as I believe that worms, if left in the ground, are far more valuable than all the fish which might be taken. My own fishing experiences incline me to the view that a lobworm will be accepted by a fish when other worms are refused, and furthermore that the angler who carries lobworms in his bag will find his requirements fewer than if he had a mixed bag. On more than one occasion I have taken two or even three trout with the same lobworm, much to the discomfiture of friends who were unable to interest a fish with their nondescript collection.

It is my opinion that in the garden, as long as the supply of food lasts, the worm will be content to remain, but when it can no longer reach any food without withdrawing its tail it will move to other quarters. To do this it burrows laterally, coming to the surface to feed at some little distance from its previous position, moving again as occasion demands but always leaving behind the open tunnel it has vacated, which then carries off any surplus surface water. Obviously, if we want to have a number of worms working for any length of time on one particular piece of ground, they must be fed, otherwise they must be expected to seek their food elsewhere. But to feed worms is easily accomplished. All will be well, if we make certain that there is some decayed or decaying vegetable matter within easy reach. The earthworms will then consume all they require and leave behind their casts - the very quintessence of humus - which in turn are digested by the living plants.

Not all that is found in the casts, however, is of plant origin: much of it may be fine particles of earth brought up from the sub-soil. Earthworms, during the hours of darkness, draw a considerable amount of food into their tunnels to be consumed in the daytime; this is always liable to become mixed with the fine soil particles. The mixture of food and soil, passing through the stomach of the worm, is later deposited mostly on the surface, but some, as explained earlier, is left in the tunnels. Ripe compost spread over the ground will keep this unpaid labour force in good health, and that is the reason why compost is so valuable - I would go so far as to say vital - in vegetable cultivation. If we agree that worms are necessary, then the question of their well-being is one of supreme importance and must not be overlooked. The earthworm must feed and it is our obligation to see that its food supply is suitable and adequate.

The Earthworm's Food

In the days when farmyard manure was plentiful this problem of supply was easily solved, but now that natural manure is scarce the earthworm has, in many cases, to attempt to adjust its diet to consist chiefly of chemical products. How it has fared on its changed diet can be observed in any garden where nitrate of soda, sulphate of ammonia, superphosphate, or any of the potassic concoctions have been regularly used. In my own experience the first indication that all was not well with such treatment of the soil was the disappearance of the lobworm, which quickly showed its disapproval of a chemical diet. The worms which did remain were by comparison in every way inferior. They were usually to be seen tied up in a knot, looking half dead - as indeed they were - and in this condition could not be expected to work. They became drones, mere occupants of the ground. Some were nearly white and others, with dark heads, looked more like miniature snakes. Where such conditions prevail the search for a good lobworm will be long and wearing, for the majority will have quickly disappeared in quest of their natural food - decaying vegetable matter. The lobworm, at least, will never be misled by all the advertised claims of artificial manures.

Soil Assessors

That the lobworm will always be found in the richest soil is proved by some observations made last year. It was decided to grow a crop of cucumbers in a greenhouse which had proved unsuitable for tomatoes. There where hot-water pipes running parallel to the central path on each side, and about two feet from the outer wall. The space between the pipes and the wall was filled with compost, which was then unripe, it being my intention to let it ripen inside. The space between the pipes and the path received a dressing of very old compost. I made eight boxes, four for each side, which were 4 ft. long by 2ft. wide and 2ft. high; into them, after filling them with good compost, I put my cucumber plants.

At the very end of the season when the boxes were taken away one very interesting fact emerged concerning the wood which had been used. The boxes had been made from new wood which had only a few months before been sawn from trees grown on the estate. First a board had been placed along each side of the path and the four boxes were then evenly spaced and joined to this edging board, the same wood being used throughout. When everything was dismantled it was found that where the boxes had been filled with good compost, the wood was in a very advanced stage of decay,

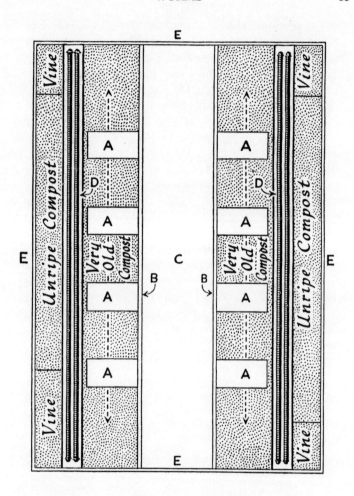

PLAN OF GREENHOUSE

A = Box, 4ft. x 2ft. x 2ft. high. B = Edging board.
C = Central path. D = Hot water pipes. E = Outer wall.

but the portions which had merely formed an edging to the path were perfectly sound. This showed that the active bacteria in the good compost were able to make an impression on the new wood, but where the older compost had touched it there was no visible effect. Here was one indication that little risk is run in adding woody material to the compost heap.

But the behaviour of the worms was even more interesting. When the border in front of the pipes was dug over, lobworms were observed in quantity where the boxes of good compost had stood, but not one was found where the spaces had been, although the distance was no more than a few feet. I did not need any help in assessing the value of the soil; the presence or absence of the lobworm confirmed my view in no uncertain manner.

This is just one instance among many of the place of the earthworm in indicating the degree of fertility of any given piece of ground. If it is sound practice to judge the quality of a farm by the condition of its livestock, may we not be equally justified in assessing the fertility of the soil from observation of its population in the shape of its worms?

Chapter 9

CLOCHES

Although I have not had very many years experience of cultivation under cloches, I think a number of gardeners may derive benefit from a statement of some of the considerations which the use of cloches has brought to my mind. I believe these miniature greenhouses - as they may well be called - have come to stay, and as time goes on they are likely to be more highly valued than they are at present.

Bridging the Gap

It seems to me that we have adopted the greenhouse proper just a little too soon. The step from the cold frame or the old fashioned hot-bed, to the greenhouse, which we have for many years now regarded as essential, involves too great a change, and a sounder practice is to transfer from frame to cloche and then, if necessary, to the greenhouse.

The greenhouse, in my experience, has often proved a liability rather than the asset so many cultivators believe it to be. Admittedly the limitations of a cold frame are soon reached, and many uses can be found for the greenhouse which would be impossible with a cold frame, even if the frame were placed on a good, old-fashioned hot-bed. The raising of early vegetable seedlings is much more comfortably undertaken in a greenhouse, but have we, perhaps, at times

allowed comfort to overide other considerations? Comfort in
gardening operations, and the best results may not always be
compatible and apart from its comfort, what else has the
greenhouse to offer? Some gardeners may claim that a host
of plants can be cultivated in a greenhouse which could not
possibly be grown in a cold frame or even with the help of a
hot-bed. I agree, but if we make a clear distinction between
essentials and luxuries, we immediately narrow the gulf
between these two aids to horticulture. There are few of
the ordinary flower or vegetable seeds which cannot be
successfully germinated and grown in a frame placed upon a
hot-bed and it is my belief that far too many gardeners, who
have all the material necessary for the formation of a hot-bed,
raise their seedlings on the greenhouse shelf merely because it
has become their practice to do so.

Greenhouse Cultivation

The usual procedure is to make up shallow boxes with a
special mixture of soil in which to germinate the seed and
bring on the seedlings. In order to make success doubly sure
the soil is sterilised either by steaming or baking, or by the
addition of one of the special substances which are sold for
this purpose. This is, no doubt, a satisfactory system for the
commercial grower who annually raises thousands of plants
more than he requires for his own use, but the private
gardener who grows only those plants which are sufficient for
his needs may find disadvantages in such a practice.
Probably the rest of the garden suffers from neglect while
seedlings are raised in the greenhouse. Fine, healthy, sturdy
plants may be grown in this way, but how often is proper
attention given to the soil in which they are finally to be
planted? Too often, I am afraid, the grower is so satisfied
with the appearance of his seedlings, raised in the greenhouse
in specially prepared soil, that he gives insufficient care and
attention to the condition of the ground outside, and believes

that his plants are so robust and well developed that with reason they can be expected to grow in almost any kind of soil. Thus we see that, although seedlings may easily and comfortably be raised in a greenhouse, we may be in danger, at the same time, of neglecting to enrich the plot in which the young plants will be expected, ultimately, to reach maturity.

In former days we were very much more dependent upon the hot-bed, into the formation of which went a large quantity of autumn leaves, waste vegetable matter, and a certain amount of horse manure. This generated the heat required for the germination of our seeds. After the raising of the seedlings, a few vegetable marrow plants could be put in for a summer crop, and when the hot-bed (which would only serve for one season) had to be dismantled, the vegetable garden benefited because the material would be spread over the ground. In this way the general fertility of the garden was being maintained by the application of the material used in the raising of the seedlings.

The Importance of Soil Fertility

But do what we may to lessen the labour of growing seedlings in greenhouses, I see no signs of a comparable effort being made to restore fertility to the soil outside. This is a question which is being universally ignored, but the adoption of the cloche as an integral part of garden equipment would be a wise move to make before we deplete our diminishing store of humus any further. It can truly be said that 'cloches catch the sunbeams', but even their most ardent advocates can hardly claim for them that they can 'catch' humus. Without the humus, cultivation under cloches will never be really successful: soil fertility is the first essential. If cloches are to be introduced into any garden, the soil itself must be rich. If they are used on poor soil, they will inevitably prove a failure - just as inevitably as their success is assured in a garden which has been regularly enriched with compost.

The Simplicity of Cloches

I do not think any person will find these miniature greenhouses difficult to understand or to use. Gardeners are accustomed to growing their plants in rows, and the continuous cloches are laid out to cover such rows. According to the height and size of the cloches used, many plants can remain under their covering until the crops mature. Seedlings of most vegetables can be raised under cloches with excellent results. The soil under the cloches must be in good heart, otherwise strong, healthy plants can never be produced. Therefore our first consideration, before sowing the seed, must, once again, be the humus content of the ground.

Chapter 10

SCIENCE IN HORTICULTURE

Years ago a phrase was coined which ran something like this, "Practice is the gardener's path and science the light to lighten the path". What a satisfactory thought for the scientist and what a comforting one for the simple gardener! The picture presented is that of a man groping in darkness along a path, when gradually a light is brought to bear on what had previously been an obscure and dreary way, and the simple soul is enabled to make better speed on his journey. It is taken for granted that the torch-bearer would lead the way. So far so good, but in the case of science as applied to practical horticulture I think we ought to examine this picture a little more closely.

The Soundness of Practice

To begin with let us get the case for practice and science right. Who said that practice needed a light? Not practice but science. Practice is a native, treading old familiar paths, never at a loss in the matter of direction with a wealth of reason for her beacon and capable of finding the way under any and every condition. Practice, in gardening, never leaves the soil and, just as long as she follows no other guide than reason or common sense, is always sure of her ground. These two companions - reason and common sense - are all that practice has been accustomed to and have always been her guiding light. Practice has always maintained that there is no short cut to success where nature is concerned and has established this truth after years of patient toil. In gardening, practice knows that pests and diseases are always liable to appear, but has been taught to believe they are there for a purpose.

Practical gardeners are convinced that all gardening operations must be based on nature, whose laws must be obeyed if success is to reward their efforts. So-called science is too often led away from the laws of nature and substitutes principles of her own, which nature would reject.

Farmers are only too well aware of the increase of disease in their livestock, but my interest as a practical gardener is, of course, centred on plants. Members of my profession agree that there has been a large increase in the number of pests and diseases in plants during the last hundred years. They acknowledge also that on this account the cultivation of the common garden vegetable has been made both more difficult and also much more costly. The incidence of pests and diseases is bound up with the problem of soil fertility, and it is in this direction that science has so lamentably failed.

Science has at least shown how easy the exploitation of the good earth may be: this is a valuable warning for which the practical gardener is grateful. But the scientist and the practical gardener have, as it were, looked at the picture from opposite sides, and the achievements of science have only convinced the cultivator more strongly than ever that any departure from the observance of nature's law is fraught with peril. If science is to be a true guide to illuminate the path of the gardener, the path chosen must conform to the laws of nature and not merely to man-made principles. Under such conditions the light of science cannot but be a blessing, and on such a path practice and science may well journey hand in hand and establish an interchange of views, opinions, and knowledge such as will make possible real progress to the benefit of all mankind.

But science, it would seem, has been too full of her own importance to try and find a use for anything not of her own creation, and has concentrated most of her time on simply destroying everything she deems harmful. So far her successes have been few, in fact it can safely be said that she has created more pests and diseases than she will ever be able to control. Science has no sense of proportion and has set out to create, if not a new heaven, then at least a new earth without any constructive plan to work from. The desire for control has outstripped her sense of the balance of nature. Anything which flies, creeps, or crawls is looked upon as dangerous. All forms of plant life which science cannot make use of must be ruthlessly destroyed, so that the flying, creeping and crawling things of nature can find no nourishment should they escape the poison sprays and other lethal weapons turned upon them.

The fact that the leaders of scientific thought have not seen fit to curb the enthusiasm of their disciples in subordinate departments can only indicate that in their opinion no caution was required. All responsibility, therefore, for the ill effects of her doctrines must be laid at the doors of science as a whole. Though it may be largely the enthusiasm of the disciple in following his own particular narrow track, that has isolated practice from science and created a gulf between the two, the result has been that science, forgetting the basic laws of nature and working from the unstable foundation of man-made principles, has brought about a widespread loss of soil fertility and this, once created, has inevitably widened the gulf. If science could be induced to give heed to the repeated warnings of both plants and cultivators that her system is at fault, a much closer relationship between gardeners and scientists could be established to their mutual benefit. Once science becomes steeped in practice and cultivates a profound knowledge of nature's law she will find practical men willing and anxious to follow her directions.

The value of any system can only be rightly judged by the result of its application to the subject in question, and on this count the scientific approach to the question of soil fertility must be considered unsuccessful in so far as it has failed to arrest the spread of disease, and the decline of soil fertility has continued unchecked.

The Divorce of Science from nature

Wherever science has held sway the results are apparent to all who care to look for them. The old order has changed and always for the worse as far as the gardener is concerned. Her presence has been felt in departments other than the soil and the garden staff of the future are likely to be a mixture of mechanics and chemists. The gardener's mind is drawn away from the study of plants to follow the life history of pests and diseases. The appearance of a new symptom of disease is hailed with delight instead of being regarded as a reproach to the gardening fraternity. Science sits in solemn session and in due course publishes her report, which is usually to the effect that after exhaustive tests a new powder has been found to control the new dreaded scourge. It seems strange to me that the magic powder only appears to exist in parts of the world which are thousands of miles from this country. Perhaps the same maxim applies to plants as to human beings - the more fashionable the complaint the more expensive the cure.

Science in gardening is for ever ignoring the rights of the soil and substituting something of her own creation which is utterly incapable of competing with nature, and science will never cease to be engaged in trying to rectify her own mistakes until she learns to do first things first.

The Soil - Our Teacher

The soil is the starting place for all horticultural knowledge. It is not a medium for experiment - nature conducted all the necessary experiments long before the first generation of scientists were born - but the soil is well worth all our study not so much for the purpose of improving it, but rather with the object of finding out what it can teach. Every day of our lives nature reveals the way to those who will take the trouble to look for it, but in far too many cases science is convinced that nature must be wrong and interferes unnecessarily.

The science of artificial manuring has become the curse of gardening. Because a particular product of manufacture appears to give good results, the scientific methods of artificial manuring are heralded as a great advance. Because a shortage of one substance is found in a certain area, that must be made good before our plants can grow well; but because in nearly all parts of the world there is shown, according to man's measurements, to be a shortage of phosphates, we are told that nature is wrong and it is our duty to rectify her error at once by dissolving some rock with sulphuric acid. If the extra amount of phosphates were really needed by the plants, nature would most certainly have seen that their needs were supplied without the assistance of man.

Nature is generous in all her gifts, otherwise how could provision have been made for the enormous increase of the earth's population which has taken place since dawn of day? Most of our pre-war troubles arose from the fact that the earth produced more than we could use. On the one hand, we had a section of the world stimulating the soil with chemical fertilisers, while another section was burning the crops. Millions of acres are now incapable of producing a crop at all; and science, on the grounds that she made possible the use of stimulants, must shoulder the responsibility. If science has created a monster she cannot control, she has only herself to blame. Science has called herself a leader, but in the eyes of the true gardeners she has been a leader without training. Nature will never willingly submit to be measured by the yardstick of science, nor will she alter her laws to suit man. If science is truly to assist the gardener, the first essential is to find out what the gardener wants and then to work out a solution of his problems from the basis of a healthy soil. There is little evidence to support the view that pests and diseases are at all prevalent in plants grown on land in good heart, land which has been maintained in health and fertility by following nature's law - the law of return. The scientist's law of the minimum is wrong and must be put aside in favour of the law of return to the soil of all waste products of the garden, field, and farm. Only this will prove adequate, and good results will quickly follow.

I have pointed out in preceding chapters how control of pests and diseases is easily and cheaply obtained by the use of compost. My own experience has abundantly confirmed that chemical manures must be followed by poison sprays, but that natural methods of replenishing plant foods always give health to the crop. No poison sprays are needed to protect the plants once the soil is made healthy, and a crop which has never required this form of protection must surely be more fit for human consumption than one which, from its infancy in many cases, has had its foliage covered with a film of poison. Unless we remove the cause of the infestation, the use of the poison spray is but a palliative measure and definitely harmful to the soil population. The feeding place of the earthworm is chiefly above ground and therefore it is safe to assume that many worms must perish where they are compelled to feed on the poisoned foliage of weeds, or through the action of the poison coming into contact with their bodies while they are searching for food. Such reduction of the earthworm population is a serious loss to the soil, and the use of poison sprays, for this one reason alone cannot be too strongly condemned. Worms are more capable of creating fertility than any chemical manures. Let me give two instances from my experience at Levens of how worms alone have rejuvenated barren soil.

An area of parkland was stripped of turf down to the loose rock, the turf being required for potting purposes. Ten years' later the stripped area was completely covered with turf without any human assistance whatsoever. The worms which were left in the loose rock had brought up to the surface their casts; the winds brought seeds from the surrounding grass which, germinating in the casts, succeeded in growing and covering the bare ground in this short period of time. After a few years no depression could be found to mark the place from which the turf had once been taken.

In the vegetable garden a hard walk of about 100 yards in length was surfaced with local material just before the war. Scarcity of labour prevented any cleaning of the path for three years and in that short period it became covered to a depth of two inches with grass and other vegetation. The surfacing material was quarried locally, spread out on the path immediately, and could reasonably be considered sterile, yet from the uncompromising material the worms formed fertility. Only by much hard labour, can man equal the results obtained by the earthworm.

A New Line of Research

Let our future research workers start their line of inquiry from the basis of a healthy soil: let them concentrate on methods, such as nature herself ordains, of keeping the land in good heart and then see if they cannot reduce to insignificance such pests and diseases as need control.

Let opposition play its part in finding the solution and let the scientist bring his genius to focus upon the understanding of nature's laws, instead, as for so long been the case, of only solving one problem by creating a bigger one. It is time to stop the isolation of each particular problem - to stop, as it were, answering the question after it has been wrenched from its context; by all means let the expert in the relevant field of inquiry bring his special knowledge to bear upon the problem, but let him view it against the complete background of nature and from the standpoint of soil fertility. No one would welcome a departure of this kind more whole-heartedly than the practical gardener.

Chapter 11

FLOWER SHOWS

I have no wish to belittle the admitted educational value of the flower show, but rather to offer a few suggestions to the ordinary visitor. His reasons for attending the show are first, to see what there is to be seen, and secondly, to make a note of certain varieties of fruit, flowers, and vegetables. Both exhibitor and visitor alike will benefit where our first reason is concerned: but only the former will draw much profit from our second purpose.

Special Strains

What the average visitor fails to realize is that those wonderful and perfectly staged exhibits, representing weeks of hard work by the exhibitor, are in many cases beyond his reach, because the exhibitor has been working on strains of flowers and vegetables which are not procurable through the ordinary channels.

Although at many shows it is obligatory on the part of the exhibitor to label each exhibit, usually a clause to the effect that a wrongly named exhibit will not be disqualified is, for this very reason, inserted. This is an excellent provision from the exhibitor's standpoint, for he can then display samples of his own specially selected strains under any convenient label. The exhibitor knows the whys and wherefores of his actions, but not so the unsuspecting visitor. He fills his notebook with a list of names, chiefly first-prize winners, and his next order for seeds can be expected to contain a number of the names he collected at the show.

But I have seen typical pot leeks, blanched just beyond the limits allowed for this type and bearing a label, 'Leek, *The Lyon*', carry off first prize in a class for trench leeks. I have seen *Dobbie's Champion* with a fifteen-inch blanch take first prize in a similar class. What the visitor seldom knows is that if he buys seed of *The Lyon* and cultivates his plants in the proper way, his *Lyon* will be blanched for a length of anything from fourteen to eighteen inches instead of the short, sturdy seven or eight inches he admired at the show, while if he grows *Dobbie's Champion* the blanch, instead of extending for fifteen inches, will most probably finish some inches short of that length. He will then either blame the seedsman for supplying wrong seeds or himself for having wrongly labelled his plant, though in reality the discrepancy is traceable to the prize-winning exhibitor. Such things happen at most shows, not because it is the intention of the exhibitor to lead the visitor astray, but merely to safeguard himself from the rivalry of other competitors.

As long as the exhibitor uses his knowledge of varieties wisely, even his expert rival will be kept in ignorance of the true facts and will attribute the success achieved to better cultivation, especially if the prize-winning exhibit bears the same name on its label as his own. In the majority of cases the visitor would be well advised to ignore the prize-winning entries and examine exhibits which the judges have passed by. They are more likely to be genuinely named and to be of the ordinary commercial varieties, which will give very satisfactory results if well grown. Only the regular exhibitor knows to what lengths a competitor will go in order to procure the same strain of seeds as his rival, for these specially selected strains under good cultivation will almost invariably be rewarded with a first prize.

The most anxious time for a successful exhibitor is not the period when the judges are at work, for before the judging begins the regular competitor can assess accurately enough the status of his entry, but the time when the show closes. If he is unlucky enough to lose specimens of his own particular strains, this may well be the work of a rival who would then be in a better position to compete with him in the future. Any person who regularly secures more than his share of first prizes will blandly tell even a stranger where he purchases his seed in order to allay any suspicion that he may be in possession of a superior strain that he has selected and re-selected for many years. It is not wise for a keen amateur to seek advice from the successful exhibitor in the matter of varieties. Even if he is your best friend, he is hardly likely to let you into his secrets or give you seed of his special strains. There is too much danger of his rivals getting hold of the strain if the stock goes outside his own control. Sooner or later the exhibitor is bound to become a plant breeder, if he hopes to achieve anything worth while on the show bench. This applies especially to vegetables.

Misleading Quality in Exhibition Specimens

In the classes for flowers the visitor is often attracted by colour, but what he seldom considers is the habit of the plant. He admires a dahlia, specially dressed for the show, and thinks how fine a plant of that description would look in his own garden. What he does not know is the amount of time spent on the cultivation of the prize-winning specimens and how few of the many flowers which the exhibitor's plants produced were good enough to be placed in his one vase. Little does he know of the many hours spent with tweezers dressing the florets of many of the truly magnificent chrysanthemum blooms.

Again, when he examines the fruit, how often does the visitor realise, when admiring a choice dish of apples, that these may not have been grown out of doors at all, but in an orchard-house. Instances without number could be recorded where a totally wrong impression is given at a show of the possibilities of the various fruits, flowers, and vegetables on view. Let the ordinary visitor be content to admire the fine exhibits, but as a companion let him have by his side a practical gardener who can often show him points which would otherwise be over-looked.

Let me give just one illustration. Some years ago I was interested in a large provincial show, at which I staged a number of exhibits. In order to be able to enter in the following potato classes,

(1)	Six Kidney	(4)	Three Round
(2)	Six Round	(5)	One Kidney
(3)	Three Kidney	(6)	One Round

I found it necessary to grow some forty different varieties, and even so it was far from easy to find sufficient suitable tubers. I have watched visitors noting down the names of some of my varieties which I knew would be entirely useless from the point of view of cropping quality, but which were purely and simply exhibition varieties. I was not, for this purpose, interested in the number of potatoes to a root, but only in the shape of the tubers. The one consideration was that every dish should be as near perfection as possible, even though it might mean lifting a row of potatoes 100 yards long in order to obtain half a dozen suitable tubers. Some of the best shaped potatoes are notoriously poor croppers and are not always of better flavour than the ordinary kinds.

Appearance or Usefulness?

Whether a show is good or bad for the district in which it is held is a very debatable point. Apart from the question of any unfairness on the part of a few exhibitors, even the honest ones will find it difficult to abstain from growing purely exhibition varieties which seldom equal the high standard of the ordinary utility types. If one intends to show produce, it is only natural to hope for a prize and concentrate on a prize-winning variety, but the type which is most likely to achieve success at the show is not always the most useful in the garden.

Chapter 12

IS DIGGING NECESSARY?

To be in a position to introduce a specific and alternative system of soil management successfully, its value must be considerably higher than that of any existing system if there is to be any hope of, or confidence in, its being universally adopted. It is not difficult to prove that the no-digging technique has important advantages over annual digging, such, for example, as economy in labour, simple effective control of weeds, the securing of early maturing of crops (with increased resistance to plant diseases and pests), and all achieved by the use of less organic matter than would be the case with the usual annual digging system.

For more than thirty years, I have had under careful observation the effect upon soil and crops in the same garden where intermittent trials have been carried out. During this time I have reached the conclusion that regular digging invariably lengthens the time needed for crops to reach maturity, that diseases and pest are more prevalent and that more weeds are propagated than are destroyed.

Whilst resting the spade will not completely eliminate the necessity for rotation of cropping, when successive applications of properly made compost to the un-dug surface of the soil has formed upon it an organic skin, then the rotation of crops ceases to be regarded with as much importance as it has commanded beforehand. Not that I would recommend the growing of any particular crop on the same plot of ground repeatedly; nevertheless none of the reasons for moving crops around the garden has ever sounded convincing to me. Surely when dealing with a small garden there is a limit to the distance apart from year to year that one can grow potatoes or brassicas or indeed any other crops which in the past may have suffered from blight, clubfoot or any other misfortune. Obviously what is needed in the case of an infested garden, is not merely a change of site as a means of protection against future outbreaks of disease but rather a review of the management of the soil, and to take immediate steps which will lead to considerable improvement in its fertility. If we ensure the creation of an organic skin over the whole of the garden and proceed to preserve the unity of the soil by ceasing to invert it annually, then such action will afford the most reliable safeguard against future attacks. My own simple remedy for defeating blight on potatoes is not by the use of sprays but to grow this crop with the minimum of soil disturbance and to plant maincrop varieties early, following with early maturing varieties after all the maincrop varieties are set. As early as possible during March I have found to be the best time to set maincrop varieties so that the haulm is matured and tough during late July — the time of year when blight spores are being carried about in the air. Not only does this ensure virtual immunity from this fungoid disease but by early planting the crop yield is appreciably increased.

Some years ago I inspected a large garden in the Huddersfield district where clubroot in brassicas was rampant. This garden was attached to a large school where the pupils were fed on the produce from the garden. At the time of my visit cabbages and sprouts were in a very unhealthy condition, the children refusing to eat the cabbages sent into the school kitchen. I recommended the use of a properly made compost that could be used as a mulch during the autumn and also to cease digging the soil. A few years later the gardener called upon me to say that he had completely cleared the garden of clubroot and the children were eating cabbages three times a week. As part of the cure I advised that all diseased material be composted and returned to the soil when properly ripe. I had already proved the value of this treatment many years ago. I have no hesitation in recommending the adoption of the no-digging system and the composting of all diseased material as a sure means of eradicating clubroot disease in brassicas.

No-digging is also an excellent means of reducing the amount of slug damage in the garden, although there are still horticultural experts who advocate deep digging for this purpose. To illustrate this point, I shall recall the time when one of the gardens at Levens was so slug ridden that six men and boys armed with darning needles could each pick up 1,000 slugs every morning, yet this daily slaughter did not save the crops from destruction. How numerous the slugs and how devastating their attack, can be realised when not a single leaf was left in this garden of two acres by the middle of June, 1924. The slugs completely devoured 10,000 seedling onions planted out in April, in less than a week; brassica seeds germinated but were eaten at ground level, while more mature plants of the family were left with nothing but bare stalks. When everything that was green had been demolished the slugs turned their attention to the gooseberry bushes, climbing the stems and eating all the leaves. Many proprietary brands of slug poison were tried but all to no avail and not until later in the year was it possible to plant any crops at all. This particular season was the worst I had ever encountered so far as slug damage was concerned, but it is a fact that not until annual digging ceased in this garden, have crops grown unmolested by these most voracious creatures.

I believe that annual digging is responsible for the spread of many pests, especially the sucking insect types such as aphids. I recently carried out a small scale trial with a dozen plum trees growing against a west wall; the border in which half the number of the trees was planted was forked over, while the border in which the remaining half dozen trees was planted was left undisturbed. The trees growing in the border that had been forked were all heavily infested with plum aphids, whilst the rest were quite immune.

On about one hundred yards of the same wall are planted a quantity of pear trees of different varieties; when this border soil used to be forked every year the trees suffered from attacks by the pear leaf blister mite to such an extent that by August the trees were completely defoliated. Now that this border is no longer forked, only one or two trees are affected, though very slightly in comparison with former years; no sprays of any kind have been used on these trees since digging or forking ceased.

In front of the pear and plum trees trained on the same wall as above I had planted in 1939 a quantity of rose bushes collected from other parts of the gardens, so that during the period of the war we could grow vegetables in the beds the roses had previously occupied. Two years ago (1949) a 'sceptic' wrote to me asking permission to visit the gardens, his purpose being to ascertain whether or not no-digging controlled greenfly on roses; he came and gave these ancient bushes a careful examination but could find only one or two tips in any way affected.

In 1946 a glasshouse was erected for the purpose of growing tomatoes. On completion of the building, the border soil was dug and later planted with tomatoes. The plants, as growth proceeded, became infested with white fly which reduced the crop considerably. In the autumn the old plants were removed but the house was not washed down before the next crop of tomatoes was planted. This time the border soil was not disturbed and no white fly appeared at any time during this year, nor has there been any signs in subsequent years when following the same no-digging technique.

A group of gardeners visiting these gardens during 1946

pleaded with me to use a fumigant to destroy the white fly pest, suggesting that since the gardens were open to the public, a tomato house full of white fly would prove a poor advertisement for Organic principles; no fumigant, however was used nor has any been needed when the tomatoes have been planted in a border where the soil has not been disturbed.

I must confess, nevertheless, that so far, I have not been successful in controlling leaf curl on peaches simply by not inverting the soil. It may be that trees grafted on certain rootstocks are more susceptible than others but of this I have, as yet, no proof. I have proof, however, that if it is possible to spray the trees with clean water each day from the time the flowers appear until the fruit begins to ripen, then leaf curl can be effectively controlled.

Up until now, I have dealt only with attacks by diseases and pests, but the benefits to be derived from no-digging extends in many other directions. By not digging we allow the earthworms all the liberty they need to work out a system for aerating the soil, which is much superior to the aerating it receives by annual digging. The layer of compost necessary to successfully follow this technique of no-digging, must be applied to the soil each year until a reasonable depth of organic matter is formed on the surface. This provides sufficient food and cover for these expert tunnellers. In return for the food provided, which has been placed in the position most favourable to their needs, they will not only increase in numbers enormously but continue their essential work nearly the whole year round. At regular intervals they will deposit on the surface of the soil their valuable casts, so that, whilst they aerate the soil in a most unobtrusive manner, they also manure its surface. Sufficient air will penetrate the ground for the needs of all plants so long as there are plenty of worms to tunnel into the subsoil. They will bury part of the organic matter found on the surface and return this in the altered form of worm casts, making a not inconsiderable contribution to the deepening of the organic skin on the soil surface. Because worms never work as freely in soils that are inverted, a substantial gain is secured by 'not digging' in this connection also.

Labour-Saving

It is a pity, to my mind, that more of the sceptics do not make it their business to visit gardens where a no-digging system is being regularly practised, for I think such visits would provide sufficient evidence to dispel all their doubts and fears completely. They would find among the many claims put forward by exponents of the no-digging technique, that of labour-saving is abundantly demonstrated, and this without any decrease in the value of the crops.

During the thirty years that I have directed operations here at Levens, digging, trenching, maturing, stimulating growth with chemical fertilisers and applying insecticide sprays and powders, all of these have, in turn, been practised. Yet with the abandonment of annual digging and by using properly made compost in the autumn, none of the above mentioned operations are now necessary. This amounts to a very considerable saving of labour, that is to say, quite apart from the saving in expense by dispensing with the use of chemical fertilisers, sprays and powders, and also the time which would be consumed in applying all of these. There is, in addition, a considerable saving in the time which previously would have been spent in the inevitable hoeing and weeding associated with a normal digging system.

Since I prefer to use undercomposted prunings, rather than have greenfly on my roses, I leave plenty of prunings to rot on the surface of the soil. And since I have no fear of spreading fungal diseases by leaving prunings of fruit trees on the ground beneath the parent trees, they are left where they fall. Then less than the usual amount of compost is later applied, because prunings left to rot on the ground are considered equal in fertility value to the equivalent amount of prepared compost.

I consider most of the top soil in a wood to have been derived from fallen branches rather than from fallen leaves, therefore I regard wood mould to be a more suitable name for this kind of soil. Whilst winter winds will carry the dry fallen leaves of the wood hither and thither, the heavier pieces of branches remain to be broken down by natural agencies until the stage of mould is reached. This is why I recommend that woody material should be used in compost heaps in preference to burning and adding the sterile ashes to a heap or spreading it on the soil. I do not expect such material to rot quickly, and expend its value, but rather that it should break down slowly in accordance with the laws of nature, so that all the plant nutrients it contains will finally be absorbed by the soil in the organic phase. The novice compost-maker would be well advised to incorporate some woody material in his heaps and to slightly increase the amount in each succeeding compost heap until he is using hard woody material in proportions equal to all other wastes.

It is a well established fact that where crops are growing strong and healthy, few weeds are to be found there, and I consider that crops growing in un-dug soil invariably compete most successfully with weeds in their midst.

I am satisfied that a wide variety of crops produce better results under the no-digging system than under orthodox soil treatment. I am also satisfied, that their healthy growth is a contributing factor in the control of annual weeds. I am further convinced that, provided an adequate amount of properly prepared compost is used in the correct manner and at the right time of year, crop failures due to faulty management will not arise. As such crops will resist and not invite attacks by disease organisms or pests; that no-digging can and will reduce the cost of production. Finally, that the adoption of this system offers an alternative to the drudgery inherent in orthodox practices. For, with no digging during the autumn and winter months, more time can be given to the collection of organic wastes and making of compost which, when applied to the ground, further reduces labour in other directions. Indeed, soil management under this system results in the formation of a soil surface so vastly improved by comparison with that produced by following the digging technique that seed sowing and planting become an ever-increasing pleasure, as seeds germinate and seedlings progress at an infinitely higher rate than under any other system.

To follow as closely as possible the example of nature entails far less work in the running of a garden and cuts down expenditure by half. Such operations as seed sowing and planting are easily performed, while hoeing and weeding have long since ceased to be a major operation, indeed the weeds, after reaching a profitable size, afford useful material for the compost heap.

A Means of Weed Control

Perhaps in the direction of weed control no-digging achieves its most spectacular success and this is all the more arresting since it is digging that has been given credit for this in the past.

The claims for digging with respect to the control of weeds rests upon a very dubious foundation. After many centuries of annual digging by countless numbers of gardeners, professional and amateur, the effectiveness of this practice would appear to be denied by the regular appearance of the familiar annual weeds throughout the whole of the year. That a garden left unattended for a short time during the early months of the year quickly becomes overgrown with a wide variety of weeds is a demonstrable fact. The only remedy at the disposal of the orthodox gardener is the free and continuous use of a hoe or unremitting hand weeding. So previous digging would appear to have done nothing more than to increase the number of weeds which diggers claim it should reduce; moreover a fair calculation of the time needed to control weeds after digging is at least four times that taken in the actual digging itself. Indeed, gardeners of this century have become so accustomed to the sight of weeds springing up all over the garden at all times of the year that they have come to regard them as the greatest bugbear they have to suffer. They do not think of this anachronism — the appearance of plants out of place and time — as being in any way related to their own previous action, otherwise they would, naturally, modify their methods of cultivation and so avoid a condition they so much deplore. With zeal they dig the ground every year and if at the time of digging the surface carries a crop of weeds these are buried along with the manure as digging proceeds; whether or not such weeds have seeded does not appear to matter, for the gardener is only interested in putting the weeds out of sight and mind.

It is a vain hope that all buried matter will rot beneath the surface, for ripe seeds of many plants are preserved if placed well below the surface of the soil. The next time the garden is dug then most of the once buried seeds will be placed on or near the surface of the soil and with the lengthening days and warmer atmosphere the seeds are not long in germinating. In many instances they germinate long before the seeds of crops that have been sown and this means much time has to be spent hand weeding if the crop seedlings are not to suffer.

Seeds of many of the plants we call weeds are capable of surviving burial for long years. When brought to the surface they may still germinate freely as witnessed by the growth of plants on the vertical slops of bomb craters or the quantity of fresh weeds that appear after old pastures have been ploughed. So long as the soil is not inverted, poppies and charlock are absent from these pastures but after ploughing they appear in vast numbers.

It is my experience that weeds are much more easily

controlled and their appearance regulated, with less effort, when the soil is not inverted. Applied during early autumn, the compost prepared in the summer months affords a convenient introduction to the no-digging technique. From October onwards I ignore any weeds that may be present for they are unlikely to seed immediately and time can be spared for their removal when they have reached such a size as to make them valuable material for composting.

Weeding along these lines should proceed more or less intermittently until planting time next year when a more thorough clean-up of all weeds should be undertaken. The compost laid on the surface of the soil during the autumn months should by now have rotted down to a fine mould so that seed sowing or planting can be undertaken with ease. There is no need to wait for March winds or sunshine to dry the surface since a compost-covered surface is always in a fit condition to receive seeds and plants. If deep drills are drawn in which to sow seeds, then expect to find weed seeds brought to the surface, but the space between the drills should remain free from weeds for some further weeks. Remember that the hoe, whilst useful in that it will destroy many seedling weeds under favourable weather conditions, is a potentially harmful tool with respect to the roots of the crops.

Indeed, in the case of the small garden the hoe can be given a long rest alongside the spade. As apart from the risk of damage to plants, unless most carefully used, buried weed seeds will be brought to the surface to provide additional exercise in their destruction later in the season.

The first two years of no-digging are probably the most critical and trying, since, during these early years in the practice of resting the spade, weeds may persist at all times of the year and the gardener will begin to doubt his ability to restrict their appearance to definite times of the year which will best suit his convenience or purpose. Nevertheless, if he persists and perseveres he will find that he really can determine within narrow limits, the time when the weeds become available to suit his purpose, and this will represent the first step towards control. Furthermore, I am convinced that the organic gardener must realise this essential factor in successful compost gardening. He cannot afford to eliminate weeds for this would mean the loss of valuable material for his compost heap. Neither does he wish to have to remove quantities of small seedlings during busy periods since such weed seedlings have little value in a compost heap. Yet he does require at certain times of year the most abundant supply of well grown weeds as grist to his compost mill.

In many of the beds which have remained un-dug for more than ten years I have far too few weeds for my purpose, so that in this respect no-digging is succeeding too well in controlling them. As a matter of fact I like to find a fair crop of weeds during May in beds which are planted in early June. These are carefully hand-weeded and the weeds added to the compost heap before the beds, from which they have been removed, are planted. Perhaps by the end of June another crop of weeds have developed and have been removed. It is seldom that more than one weeding of the beds is necessary after planting, as by the end of July the crop plants are making such progress that later weeds are smothered. A certain number of these weeds may survive, in which case when the crop plants are removed in late autumn, the earth in the beds is not bare but covered with a fair growth of weeds which will ultimately be placed on the compost heap.

Each October this sequence is followed:- The covering of the soil with compost; the removal of some of the more developed weeds and the retention of smaller ones to develop later into useful specimens, say during April or May (taking care, incidentally, that none of these produce seeds at such an early time in the season) since this would mean more frequent weeding in the summer. A general clean-up of all weeds before the beds are planted and another clean-up before the plants meet in the rows, after which the beds need no further attention until crop residues need removing.

It is true that this system can produce all the weeds one can reasonably expect to find in any garden, but I hope I have made it abundantly clear they are at all times under deliberate and proper control. The only times when they might appear to be out of hand is during the winter and early spring months. During these times they are allowed to remain and so to develop into useful material as distinct from destroying them in the condition of immature seedlings of little value either to the garden or the compost heap.

By the time no-digging has been practised for ten years then control over weeds becomes automatic. They are allowed to remain only at certain times of the year and if due care is taken whilst gardening operations are in progress, to avoid unnecessary disturbance of the surface soil, then we know just when to expect weeds to appear and can decide, under the circumstances then existing, how best to deal with them.

Compost

As previously mentioned one can hardly think of no-digging without at the same time associating it in one's mind with the necessity for Composting. Indeed, success with the no-digging technique is quite impossible until the soil has previously received a generous application of properly made compost — say a pailful to the square yard as an initial top dressing.

The first step, then, to be taken to ensure an adequate supply of compost for use during September is to make a thorough collection of all organic wastes during the spring and summer months. The search for suitable materials must proceed unceasingly throughout the year if sufficient compost to fulfil the needs of the garden is to be made available.

It must be appreciated, incidentally, that in order to ensure that the compost produced has the maximum of fertility value, it is essential to make use of the widest variety possible of organic matter. In the collection of which it is most desirable, so far as is practicable, to include the whole plant; and here is a role in which our good friends the weeds can offer a substantial contribution.

By far the most difficult situation comes from those people who have just moved into a new house or who have only recently commenced to make compost. It is unlikely that there will be enough good compost available at the right moment to ensure success of the no-digging technique. So to such people I would suggest, digging the garden for the first year, which of course will usually produce a good crop of weeds to be composted. In the second year plan to cease digging, though if despite every effort, there is not sufficient compost to meet the requirements of the whole garden for this second year, it would be advisable to concentrate on part of the garden only, dressing this with properly made compost at the rate of one barrow-load to five square yards. This is not a generous covering, rather the bare minimum which can be applied if good results are to be obtained.

It may well happen, however, that during the course of the second and third years much more material will be available for composting. In which case the whole of the garden could be dealt with at the rate suggested above (one barrow-load to five square yards of garden). In this connection I strongly urge composters to forgo the luxury of a bonfire since, if the garden surface soil is to rival that of the woodland then the natural decay of woody material and not sterile ashes must be allowed to bring about this change.

Most people who adopt the no-digging system find the first few years critical and disturbing to their peace of mind; as they are torn between two desires; to maintain the garden in an immaculate condition continuously, whilst at the same time providing all the materials needed for the building of large compost heaps. A supply of weeds is necessary for this purpose and because they provide fertility, they are often found where crops grow vigorously though few   can be expected to survive.   The alternative is to gather supplementary supplies from elsewhere; uproot the whole plant wherever possible, otherwise much that is valuable to the compost is left behind.

Being relieved of the need to dig, the time thus saved can be employed in the search for the collection of organic materials and as the available ripe compost increases in bulk then considerable saving of labour in other directions is effected. Odd moments can be utilised for gathering compost materials which suggests that this task need never conform to a rigid timetable.

Soil Temperature

The proper temperature of the soil must play an important part in the development of plants. It may vary with the seasons and the moisture content of the soil, but under natural conditions will remain constant within narrow limits. Broadly speaking soil is less capable of maintaining a normal temperature when dry although this fact may be contrary to popular belief.

Soil becomes dry, during a prolonged absence of rainfall, through the action of strong sunlight and also because of moisture taken up by growing plants which is transpired by way of their leaves; nevertheless, not all of the water evaporated by the sun or transpired by plants is lost to the soil, for much of this is returned in the form of dew or rain; thus there is a circulation of water from the earth to the atmosphere and back again to the earth.

Judging from letters which appear from time to time in horticultural journals, the formation of dew still remains something of a mystery to the cultivator of crops, according to the wide variety of reasons that are put forward in an attempt to explain this natural phenomenon. My own opinion in this respect, is that the soil temperature plays an important part in the production of dew, especially at the stage which can be observed as a mist rising from valley land after a hot summer's day.

I cannot imagine that the popular conception of this phenomenon as being entirely due to the release from the soil of heat previously absorbed from the sun, being correct. This explanation could only be true if we were to deny the existance of heat within the soil; a proposition I am as unwilling to accept as I am to believe the soil to be simply a dead inert mass. I suggest that the evidence of active volcanoes is proof of an area of intense heat deep down in the earth, producing molten lava which pours forth at intervals from the many craters dotted over the face of the earth. If so terrific a heat as this is generated at the centre of the earth it can only escape from the centre outwards, which means of course ultimately through the earth's crust. Though only in the case of volcanoes or hot springs does it escape violently, whilst in a country such as ours the escape is less obtrusive.

If such a theory can be accepted as reasonably true, then it can be assumed that the heat radiated from the earth, on meeting the colder air of the evening, will condense to form part of the dew which falls back on the earth. Whilst this is far from being a scientific explanation of the function of dew, yet it is submitted with confidence as the hypothesis of a busy gardener possessed of an inquiring mind. The formation of dew first takes place over valley land and whilst I have seen this to form over pastures and meadows, and even over mud flats by the sea, I have yet to see this happening to the same extent over a fallow field the surface of which has recently been worked by machinery. I remember walking home across some fields in the cool of a summer's evening and on crossing two pasture fields I observed that the grass was saturated with dew. But on reaching a field of turnips which, during that day had been cultivated between the rows of the plants, I noticed that both the plants and the soil were dry. Yet in the distance the mud flats were positively steaming, although quite devoid of any sort of vegetation: why this difference between the unworked flats and the worked soil of the turnip field?

My own belief, after extended observation under varying conditions, is that the starting point of this function of dew is the warm air rising from the soil, which, when meeting the cooler air above, condenses to form visible mist. The falling dew may consist in part, as a result of this technique, and the rest by the amount which is evaporated by sunshine or transpired by plants during the daytime. This will, presumably, condense at a higher altitude but will gradually fall, to mix later with mist forming at or near ground level.

And now the reader may well enquire as to the connection of this question of the formation of dew with cultivation of the soil by the no-digging system? In the first place it will be necessary, by some means, to attempt to prove that the surface of the earth is warmed from below as well as from above. To my mind the formation of dew provides an example, for I cannot bring myself to believe that soil can absorb sufficient heat from a few hours of sunshine to make possible the formation of so dense a mist or so heavy a dewfall; and which could, under favourable conditions, continue to form for longer periods of time than the length of time the sun was shining during the previous day; all of which, I believe, suggests the presence of a continuous supply of heat rising from the soil.

There are, however, other indications familiar to gardeners, which support my contention. We know, for example, how disappointingly tomato plants grow if planted in the border of a glasshouse where the air temperature and that of the soil differ widely. This is particularly so if the border soil has recently been dug, but if the soil is compacted and adequately moistened, then growth is much improved. I have found this true, also, of lettuce seedlings planted under cloches; these invariably made better progress if planted into un-dug soil in which a generous supply of organic matter had been incorporated.

Such experiences suggest the presence of bottom heat in the soil, and all that is needed to prove such a simple fact is for a proportion of certain crops to be grown in un-dug soil, to observe the reaction of these crops to this system of soil management and then to decide from the evidence whether or not the no-digging technique fulfils expectations. Moreover, because the benefits of no-digging are cumulative, it will be necessary to persevere with this system over a period of a few years. Then it will be appreciated that successive autumn mulching with compost forms an organic skin on the surface of the soil in which seeds may be sown or seedlings transplanted with confidence and with remarkable ease under any sort of weather conditions and from which completely satisfactory results will be always obtained. There are many factors contributing towards the better growth of plants under this system of soil management, but high up in the list must be placed the benefits of a warm soil.

For those who are unable to carry out such trials, I will submit evidence brought forward by others in support of this claim for warmth in the soil:-

In Vol 1. No. I of the *Growers' Digest* (the official journal of the Scottish Fruitgrowers' Research Association) there is published an interesting table which gives a schedule of minimum temperatures recorded at the surface of adjacent plots of soil of differing types during calm radiation frost periods and is dated February 27th, 1952:-

Wet loam, compacted	26.3
Wet loam, friable to three inches depth	24.9
Dry loam	24.4
Wet Sand	20.0
Dry Sand	19.1
Wet Peat	17.6
Dry Peat	10.5

I think few gardeners would expect to see such a wide difference in temperature between a wet compacted loam and a dry peat surface. It is explained that the heat which is radiated away from the soil is partly replaced by the upward flow of heat from the lower and warmer levels of the soil; the rate at which this occurs being determined by the heat conduction qualities of the soil under review. The low temperatures of dry peat and sand reflect their poor heat conductivity, whilst a second factor is their low specific heat.

Other interesting features disclosed by this report include the somewhat startling revelations that the addition of any light open humus material such as peat or straw manure to the top few inches of the soil increases liability to frost, which, in other words, means a lowering of soil temperature through soil disturbance.

My interest in no-digging was established long before I had read the above report and was based on observations and empirical trials all of which indicated that the most effective time to apply a mulch of compost was in the autumn, if it was not to lower the normal temperature of the soil to which it was applied. Subsequent frequent and heavy rains could then be expected to compact this otherwise loose mulch and increase its power of conducting the warm air from the lower levels through the crust of the earth. This, incidentally, is not to deny the value of a compost mulch applied at any time of the year, since to cover the top few inches of soil with humus cannot be other than beneficial. However, because I attach great importance to the maintenance of the highest possible soil temperature, my application of a compost mulch was designed to interfere with this as little as possible.

We have seen how, by disturbing the soil to a depth of three inches the normal temperature of the surface is reduced and this reduction increases with the drying of the surface of the soil. For example, we dry the surface of a loamy soil to approximate in general character to that of a dry sand or peat, then a considerable reduction in the surface soil temperature follows immediately. Yet another example of how a soil loses its temperature on becoming dry is provided by the use of electrified wires buried beneath the surface of the soil. Having recently installed equipment for varying the soil temperature by means of bare wires carrying low voltage electricity, it was interesting to discover how quickly this arrangement raised the temperature of the soil. I overlooked, however, the possibility of this heat fluctuating under continuous heating, because, when the current had been on for a few days and the soil above the wires began to dry out, then a thermometer registered a reduction of heat in degrees F. Indeed, it would appear that no matter by what means the surface soil becomes drier, this change is accompanied by a lowering of its temperature. We have seen how the disturbing of the top few inches of a wet compacted loam resulted in a lower temperature reading and the same thing happened when the surface soil is electrically dried, and on watering the soil above the electric wires the previous higher temperature was at once restored.

With the above in mind we can reckon that all soils, when disturbed, will lose much of their inherent heat; also that the loss will be greater if disturbance creates a dry surface and that this loss may continue for an unspecified length of time; probably until the soil regains its original state of compactness.

It is elementary knowledge that if soil is dug in the spring for the benefit of immediate planting, this dug soil must be firmly compressed or plant roots will be unable to obtain sufficient of the nutrients required for healthy growth. Also, the soil will quickly lose moisture if left for long in a dry, loose condition. Of equal importance is the loss of bottom heat which inevitably occurs in such soils since heat ranks with moisture as an essential in the prolification of plant roots. Even the most confirmed digger cannot claim other than a lowering of the soil's temperature as a result of his labour, which cannot, by any stretch of imagination, be reckoned an advantage either to the soil or himself.

Doubtless there are times when it is advisable to secure a cool soil surface for the benefit of growing plants and especially if this can be made to serve a dual purpose.

Under the influence of strong sunshine much water is evaporated from the soil, also plants transpire much through their leaves, and unless this can be checked, then they will wilt and temporarily cease to grow. The most popular method of stopping this loss of moisture from the soil is to hoe the soil around the suffering plants in order to create a dust mulch, or alternatively to apply a mulch of dry material to act as an insulating layer between the soil and the sun. From my own experience, however, the best results from mulching are obtainable by first soaking the ground with water, then adding a cover of such material as dry sawdust, sand, peat or long strawy matter.

By previously soaking the soil, the absorbed moisture acts as a reservoir and the dry matter on the top of the soil acts as an insulator against the power of the sun; if simply a wet mulch is applied the moisture it contains might well soon be evaporated by the sun, thus shortening its period of usefulness.

Perhaps the best of all mulching material is a local stone or similar inorganic material, since, if applied to a reasonable depth, the sun is powerless to act upon the underlying soil. Indeed, market gardeners in America employ stone mulching to a far greater extent than gardeners in this country, also whilst the use of paper is practised by many in the States, both to protect the surface soil from loss of moisture through evaporation and as a means of securing quicker germination for seeds. An area to be sown with seeds is first covered with rolls of paper some days or even weeks previous to the actual sowing. The effect of this procedure being to discourage the growth of annual weeds and at the same time to make the soil warmer and moister than it would otherwise have been if left fully exposed to the heat of the sun.

If, then, we consider the positive features from the different examples described, I contend that we are able to make out a good case against the too frequent disturbance of the soil; nevertheless, the diggers rightly claim that by burying fairly deeply, compost or farmyard manure, say several inches below the surface, the roots of the plants are encouraged to search for this deeply buried plant food, and at that level of course, the moisture it contains is well protected from the evaporating powers of the sun. They claim, in effect, that by burying organic matter in October, provision is being made for the needs of their plants during the hot days of the following June, and even if there is a loss of soil heat as a result, it is claimed that this is more than compensated for by the value to the plant of this buried organic matter.

Of minor importance to the followers of digging is the fact that, under their system of soil management, plant roots are virtually compelled to travel in a downward direction in search of something which might well have been laid on the surface, for it is a significant fact that if compost is spread upon the surface on un-dug soil during October, then all but taprooting subjects will develop a natural rooting system with the feeding roots radiating from the centre in all directions and lying close to the surface. As with permanent fruit trees, so with the more surface rooting lettuce: the deeper the roots have to plunge in search of nutrients, and at levels beyond the influence of the sun, to that extent will the quality suffer.

How inferior is the fruit of an apple tree whose roots have penetrated into subsoil and how quickly the quality of the fruit is improved if the tree is lifted and the offending roots removed? The difference between two lettuces, one with the deeply penetrating roots and the other gathering nutrients from near the surface may not, in fact, be as striking as the example with respect to the apples, yet I am convinced from experience, that a difference does exit and that a lettuce which has never sent its roots down to an abnormal level in search of plant food, suits my taste much better than its opposite number.

This final chapter consists of extracts from Mr King's booklet 'Is Digging Necessary?' written in 1952.

Appendix

Original Introduction

By Sir Albert Howard, C.I.E

(Fellow of the Imperial College of Science)

Although the grow-more-food policy has fulfilled its primary object of saving this country from starvation, nevertheless in one important respect it must be regarded as an expensive failure. It has missed a unique opportunity of killing two birds with one stone, safeguarding the vital food supply and, at the same time, establishing the public health system of the future on the firm foundation of a fertile soil. In September 1939 two roads were open to the various Ministries concerned with food: (1) to cash in, by means of artificial manures, the stores of soil fertility still left to us so to transfer the soil's capital, humus, to the profit and loss account regardless of the poor quality and indifferent nutritive value of the food so produced, or (2) to set in motion a supreme national effort to restore and maintain the fertility of our soils, so that the maximum quantity of food of the highest possible quality could be produced for the men and women engaged in the war effort, and to leave the land in such a condition that an entirely new system of social services and public health could be established after the war. The first of these alternatives was by far the easier one. It was duly followed. Food was produced anywhere and anyhow: vast quantities of subsidised artificial manures were forced on the farmers: the reserves of the vital humus were sadly depleted: the seeds of the future trouble in the shape of malnutrition, inefficiency, and actual disease were well and truly sown.

But here and there in this dark picture rays of light could be observed. A few pioneers followed the right path by first of all getting the land into good heart and leaving it in much better fettle than when the drive for more food began. To this select band the author of this book belongs. Mr. King has not only got the gardens at Levens Hall in South Westmorland, which are in his charge, into excellent shape for vegetables and fruit by the faithful adoption of Nature's law of return, but he has gone much further. He has made these gardens a place of pilgrimage for the schoolmasters of Westmorland and for the many who have attended his excellent evening lectures at Kendal, Milnthorpe, and other centres. He has also recorded that portion of his experience which applies to cottage gardens and allotments in *The Compost Gardener*, published by Messrs. Titus Wilson and Son, Ltd., 28 Highgate, Kendal. This book, which appeared in November 1943, proved an immediate success and is already in its second edition. As a result, garden after garden is giving up artificials and poison sprays, and beginning to produce real food instead of artificial nourishment.

In the present book Mr. King has, with the help of my private secretary, Miss Ellinor Kirkham, carried the story of compost a stage further and has shown how far-reaching is the effect of freshly prepared humus in preserving the soil texture, in maintaining and stimulating the soil population, and thereby providing the crop with all the food materials needed without any help form applied chemistry. He has also recorded his experience of the great natural resistance to all kinds of pests, including virus diseases, which humus-filled soil confers on both vegetables and fruit trees. His experience amply confirms the view now rapidly gaining ground that the appearance of pests in a garden is a sure sign of bad staff work on the part of the gardener himself.

Perhaps the most original chapter is the one on weeds, in which these plants are not regarded as troublesome intruders, but actually welcomed in the late summer and autumn as useful green manure crops for collecting from the soil the surplus food materials, which otherwise would run to waste.

The fresh organic matter so produced provides the soil organisms in the early spring with food from which the vital combined nitrogen needed by the next crop is manufactured. Weeds must, therefore, be controlled and not destroyed, a point of view which will afford much comfort to the readers of this book.

Another chapter in which new ground has been broken deals with cloches. These miniature outdoor greenhouses are now coming into general use in this country and are certain to become much more popular once it is realised that cloches and compost must run in harness. Simple means are suggested by which the cloche can come into its own.

In recording the deleterious effects of artificial manures on some of the plots at Levens Hall, Mr. King has unconsciously indicated a new and promising field for the research workers of tomorrow. His observations point to the need for keeping a very careful watch on all land under experiment in order to determine what are the after effects of chemical fertilisers. Nowadays the yield is chiefly studied. This is not enough. What is the effect of say annual dressings of an artificial like super-phosphate on the earthworm population and on the way the soil responds to the spade? Gardeners who have to cultivate the same land year after year are in an excellent position to observe the ultimate consequences of using such manures. All may go well for a year or two, but what happens afterwards? Mr. King has much to say on this point on which the research workers of the future will do to ponder. To avoid the pitfalls which he indicates scientific investigators will have to become practical gardeners and carefully observe the final results on the soil itself of any advice they have to give. They must also secure the approval of the unseen labour force of the gardener, the soil organisms, including the earthworm, for any innovation they suggest. Had this been done in the case of manures like super-phosphate and sulphate of ammonia, it is safe to say that neither of these soil poisons would ever have emerged from the seclusion of the experimental stations.

I have no hesitation in recommending this book to all gardeners who intend to make their particular corner of England productive and who are interested in the principles underlying gardening. They will find no half-measure, and no compromise with official views, in Mr. King's book. He has based his conclusions on a close study of the processes of Nature and on a wide experience as a practical and successful gardener. In all this he has been greatly assisted by his exceptional powers of observation, by his uncompromising honesty, and by the capacity, rare among the members of his craft, of expressing his conclusions in vigorous and compelling phrase.

Further Reading

In our gardening classics series

How To Enjoy Your Weeds
by
Audrey Wynne Hatfield

This little book really is a joy for all those interested in gardening and nature. In a delightful manner, the author demonstrates a lifetime's knowledge of flowers and herbs with her detailed descriptions of each species. She includes a number of recipes for herbal drinks and country wines and also fascinating old remedies for common aches and pains. In addition to this, there are explanations of the folklore and mythology that surround many of the plants included, and the text is complemented by the author's own superb illustrations. Arranged in alphabetical order, the book is easy to follow and a pleasure to read.

Audrey Wynne Hatfiled was raised in South Yorkshire and spent much of her childhood exploring the countryside. She studied art and spent a period as an actress, before eventually taking up writing. She settled in a remote part of Hertfordshire as close to nature as possible, where, amongst other things, she grew soft fruit and herbs. Her talents are combined here to provide us with a book that would take pride of place on any bookshelf.

ISBN: 0953364607

Available through your local bookshop

Useful Addresses

The Henry Doubleday Research Assoication
Ryton Organic Gardens, Ryton-on-Dunsmore
Coventry CV8 3LG

&

Yalding Organic Gardens, Yalding, Kent ME18 6EX

The Herb Society & The National Herb Centre
Deddington Hill Farm, Warmington, Banbury
Oxfordshire OX17 1XB

Index